anatomy of exercise for
WOMEN

anatomy of exercise for
WOMEN

Editor:
Lisa Purcell

FIREFLY BOOKS

A Firefly Book

Published by Firefly Books Ltd. 2013

First printing

Publisher Cataloging-in-Publication Data (U.S.)

A CIP record for this title is available from the Library of Congress

Library and Archives Canada Cataloguing in Publication

A CIP record for this title is available from Library and Archives Canada

Published in the United States by
Firefly Books (U.S.) Inc.
P.O. Box 1338, Ellicott Station
Buffalo, New York 14205

Published in Canada by
Firefly Books Ltd.
50 Staples Ave, Unit 1
Richmond Hill, Ontario L4B 0A7

Printed in Canada

This book was developed by:
Moseley Road Inc.
123 Main Street
Irvington, New York 10533

For Moseley Road:
President: Sean Moore
General Manager: Karen Prince
Project Editor/Designer: Lisa Purcell
Production Designers: Adam Moore, Danielle Scaramuzzo, Terasa Bernard
Photographer: Jonathan Conklin Photography, Inc.
Models: Elaine Altholz, Goldie Oren, Melissa Grant
Contributing Writers: Philip Striano, DC; Dr. Abigail Ellsworth; Hollis Lance Leibman; Amy Pierce

CONTENTS

Introduction: Fit & Feminine 8

Full-Body Anatomy 12

FLEXIBILITY EXERCISES 15

    Neck Side Bend 16

    Triceps Stretch 17

    Posterior Hand Clasp 18

    Chest Stretch 20

    Swiss Ball Kneeling Lat Stretch 21

    Latissimus Dorsi Stretch 22

    Toe Touch 24

    Cat and Dog Stretch 26

    Piriformis Stretch 28

    Hip Stretch 29

    Hip-to-Thigh Stretch 30

    Spine Stretch 31

    Swiss Ball Hip Crossover 32

    Knee-to-Chest Hug 34

    Iliotibial Band Stretch 36

    Quadriceps Stretch 37

    Standing Hamstrings Stretch 38

    Standing Calf Stretch 39

    Child's Pose 40

UPPER-BODY EXERCISES 43

    Chair Dip 44

    Chair Crunch 46

    Overhead Press 48

    Alternating Chest Press 50

    Standing Fly 52

Upward Plank. 54

Swiss Ball Pullover. 56

Swiss Ball Triceps Extension. 58

Swiss Ball Fly. 60

Push-Up. 62

Prone Trunk Raise. 64

Dumbbell Upright Row. 66

Alternating Dumbbell Curl. 68

CORE-TRAINING EXERCISES 71

Crunch. 72

Half Curl. 74

Seated Russian Twist. 76

Spine Twist. 78

Oblique Roll-Down. 80

Bicycle Crunch. 82

The Boat. 84

V-Up. 86

Backward Ball Stretch. 88

Plank 90

Swiss Ball Transverse Abs. 92

Swiss Ball Rollout. 94

Foam Roller Calf Press. 96

Foam Roller Diagonal Crunch. 98

Foam Roller Supine Marches 100

Tiny Steps. 102

Double-Leg Abdominal Press 104

The Twist 106

Standing Knee Crunch 108

Power Squat. 110

CONTENTS continued

Swiss Ball Reverse Bridge Rotation 112

Swiss Ball Sitting Balance 114

Swiss Ball Hip Circles 116

Swiss Ball Reverse Bridge Roll 118

Abdominal Hip Lift 120

Leg Raise 122

LOWER-BODY EXERCISES 125

Foam Roller Iliotibial Band Release 126

Swiss Ball Jackknife 128

Shoulder Bridge 130

Foam Roller Bicycle 132

Single-Leg Circles 134

Scissors. 136

Wall Sits 138

Stiff-Legged Deadlift. 140

Forward Lunge 142

Lateral Lunge 144

Dumbbell Lunge 146

Dumbbell Calf Raise 148

Kneeling Side Lift 150

Put It All Together: Workouts. 153

Glossary. 158

Credits & Acknowledgments. 160

INTRODUCTION: FIT & FEMININE

Ever-increasing numbers of women are taking up fitness regimens, whether joining gyms, running in marathons, or simply spreading a mat in the living room and trying out some home exercise. The reasons are many: some women want to drop a few pounds and tone up their thighs to fit into those skinny jeans; others are looking to improve their overall health and increase their energy levels; some use exercise to manage stress and improve their mood; and of course, many begin working out for a combination of all those reasons. The simple fact is exercise makes us all feel better—and look better, too.

For whatever reason you've decided to follow an exercise program, you'll find plenty of valuable information and tips in the following pages. You'll find a guide to a comprehensive exercise program, devised with attention to your whole-body anatomy. The first group of exercises focuses on flexibility, demonstrating moves that can warm you up before a longer workout or just get you going to start your day. Sections on the upper body, the core, and the lower body target those areas most of us want to improve. Performed together, these exercises will not only enhance your figure, but also increase you body's performance levels.

FIT & FEMININE

Fit and feminine: the goal for most women undertaking an exercise regimen. We all want to look our best and feel our best, too, so that we can perform at peak levels, with energy to spare. With women's busy lives, it isn't always easy to fit in everything that needs to be done in a day, but making time to exercise is one of the best investments you can make in yourself.

This book is divided into five sections: the first group of exercises focuses on flexibility, demonstrating moves that can warm you up before a longer workout or just get you going to start your day. Sections on the upper body, the core, and the lower body target those areas most of us want to improve. The final section offers sample workouts. Performed together, these exercises will not only enhance your figure, but also increase you body's performance levels.

MAKE TIME FOR FITNESS

All too often we put off starting an exercise program because there simply doesn't seem to be enough time in a day. Yet, it may be easier than you think. For example, using this book as a guide, you can work at home, saving the extra time (and money) that working out at a gym demands. Just find yourself a space (the living room, for instance), and set aside just 10 to 30 minutes a day, two or three times a week. Designate a regular time (say, after dinner); this encourages you to stick to a consistent schedule. Just as you can build up the weight on your dumbbells, so too can you build up the hours per week you spend exercising. Dip into this book over time, and don't be afraid to try something new; you may find an exercise that challenges you in a new way or discover that you're weak or strong in an area you never knew existed. Pay attention to what your limits are, and then work toward exceeding them. As you get more comfortable with the workout, devote more time to it to see faster, better results.

WORKING IN NEUTRAL

Neutral spine, also known as neutral posture, is an important concept that you need to understand before you begin practicing exercising. Neutral spine is crucial for ensuring that you properly target and strengthen the muscles of the core, and it also keeps you in a more efficient position for movement.

Neutral spine is the proper alignment of the body between postural extremes. In its natural alignment, the spine is not straight. It has curves in the cervical (neck), thoracic (upper), and lumbar (lower) regions. Neutral alignment helps to cushion the spine from too much stress and strain. Controlling pelvic tilt is one way to begin helping to balance the spine. As certain muscles of the back and abdomen contract, the pelvis rotates. As the pelvis rotates backward, the lumbar curve increases. As the pelvis rotates forward, the curve of the low back straightens.

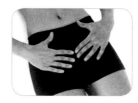

To find neutral spine while lying on your back place your thumbs on your hip bones and your fingers over the pubic bone (the bone between your legs), creating a triangle. All the bones should line up on the same plane—no tipping back or to one side should be present. The triangle should appear "flat," with all corners on the same plane. This position will prepare you for exercising when you are lying on your back.

If you are exercising on your stomach, you can find the neutral spine by pressing your pubic bone into the floor until you feel your back flatten slightly or your stomach lightly lift off the floor. Tuck your chin so that your forehead has contact with the surface, and your neck is now ready for strengthening. This position not only protects your back and your neck as you exercise, but it also allows you to exercise more productively. Maintaining neutral posture will help decrease the risk of injury and increase the efficiency of movement or exercise.

EXERCISE EXTRAS

To add variety to your fitness regimen, take advantage of everyday objects around the house: use a chair as a prop for dips and crunches, for instance, or take advantage of steps for lunges and calf exercises.

Many of the featured exercises incorporate equipment—all reasonably small tools that add variety and challenge to your workout.

Hand weights and dumbbells. Several of the toning exercises call for small hand weights or dumbbells. You can start with very light, 2-pound weights (or even lighter substitutes, such as unopened food cans or water bottles), and then work your way up to heavier ones.

Both hand weights and dumbbells add resistance, increasing the benefits of many exercises. You can use either one for any exercise that calls for a weight. If you decide to invest in a set of dumbbells, look for an adjustable model that allows you to easily vary the weight

Medicine ball. A small, weighted medicine ball, which is used like a free weight, can also be used in any exercise that calls for a hand weight.

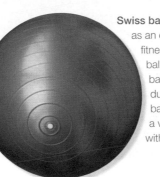

Swiss ball. Also known as an exercise ball, fitness ball, body ball, or balance ball, this heavy-duty inflatable ball is available in a variety of sizes, with diameters levels. Be sure it comes with a solid-locking mechanism that makes adding and subtracting weight disks fast and easy.

ranging from 18 to 30 inches. Be sure to find the best size for your height and weight. A Swiss ball is an excellent fitness aid that really works your core. Because it is unstable, you must constantly adjust your balance while performing a movement, which helps you improve your overall sense of balance and your flexibility.

Resistance band. Also known as "fitness band," "Thera-Band," "Dyna-Band," "stretching band," and "exercise band," this simple tool adds resistance to an exercise. You will see two types of resistance bands, one with handles and one without; both are amazing pieces of fitness equipment, which effectively tone and strengthen your entire body. Bands act in a similar way to hand weights, but unlike weights, which rely on gravity to determine the resistance, bands use constant tension—supplied by your muscles—to add resistance to your movements and improve your overall coordination.

GETTING STARTED

Although you may be tempted to dive right into your workout, warming up is essential to any exercise program. Warm-ups will increase the benefits of exercising and help decrease the potential of sustaining injury. The basic kinds of warm-ups fall into two categories: cardiovascular exercises and stretches. Cardio exercises stimulate blood and oxygen flow through your body. Try running in place, jumping rope, spinning or cycling, or even brisk walking. Stretches, such as those found in the first chapter, gradually and smoothly lengthen the muscles, maximizing their flexibility.

HOW TO USE THIS BOOK

In the step-by-step chapters of this book, you'll find photos with instructions demonstrating how to execute each exercise and some tips on what to do to perform it correctly—and what to avoid. Some exercises have accompanying variations, shown in the modification box. Alongside each exercise is a quick-read panel that lists the exercise's major target, level of difficulty, and benefits. Also included is a list of precautions: if you have one of the issues listed, it is best to avoid that exercise. Each exercise also features illustrations showing key muscles. As you work out, visualize the muscles that you are engaging—it will help you maintain optimal form.

FULL-BODY ANATOMY

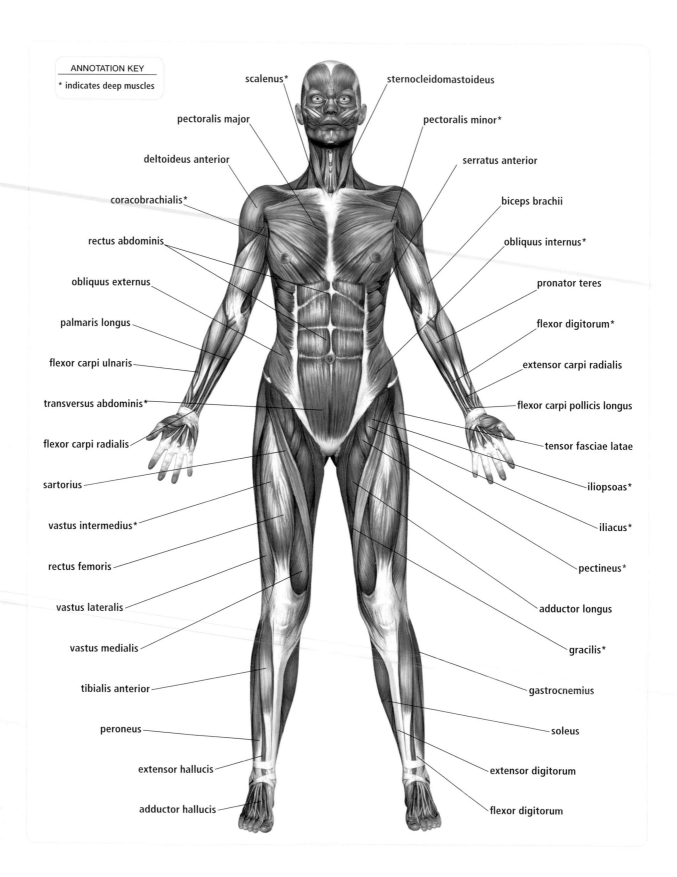

ANNOTATION KEY
* indicates deep muscles

scalenus*

sternocleidomastoideus

pectoralis major

pectoralis minor*

deltoideus anterior

serratus anterior

coracobrachialis*

biceps brachii

rectus abdominis

obliquus internus*

obliquus externus

pronator teres

palmaris longus

flexor digitorum*

flexor carpi ulnaris

extensor carpi radialis

transversus abdominis*

flexor carpi pollicis longus

flexor carpi radialis

tensor fasciae latae

sartorius

iliopsoas*

vastus intermedius*

iliacus*

rectus femoris

pectineus*

vastus lateralis

adductor longus

vastus medialis

gracilis*

tibialis anterior

gastrocnemius

peroneus

soleus

extensor hallucis

extensor digitorum

adductor hallucis

flexor digitorum

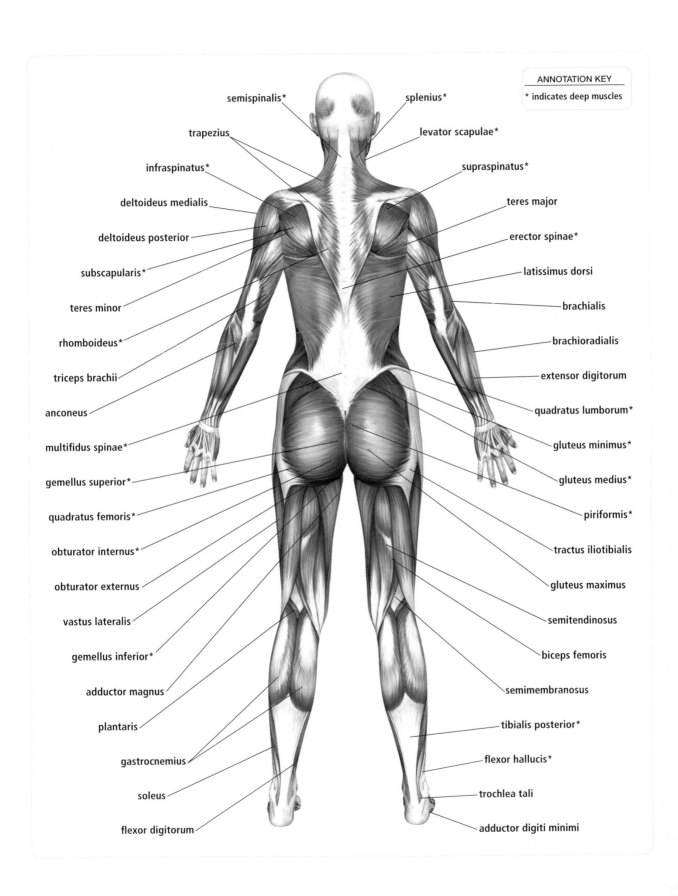

semispinalis*

splenius*

trapezius

levator scapulae*

infraspinatus*

supraspinatus*

deltoideus medialis

teres major

deltoideus posterior

erector spinae*

subscapularis*

latissimus dorsi

teres minor

brachialis

rhomboideus*

brachioradialis

triceps brachii

extensor digitorum

anconeus

quadratus lumborum*

multifidus spinae*

gluteus minimus*

gemellus superior*

gluteus medius*

quadratus femoris*

piriformis*

obturator internus*

tractus iliotibialis

obturator externus

gluteus maximus

vastus lateralis

semitendinosus

gemellus inferior*

biceps femoris

adductor magnus

semimembranosus

plantaris

tibialis posterior*

gastrocnemius

flexor hallucis*

soleus

trochlea tali

flexor digitorum

adductor digiti minimi

ANNOTATION KEY

* indicates deep muscles

13

FLEXIBILITY EXERCISES

Flexibility improves your performance when you bend, lift, and reach your way through daily life. Becoming more flexible can improve your physical performance and decrease your risk of injury. And, yes, we are all born with natural flexibility in certain areas but face challenges in others, but flexibility can always be improved over time and with practice. The following exercises all deliver targeted stretches to key muscles, and performing several before a workout is a great way to warm up. But stretching isn't just about warming up—regular stretching imparts multiple benefits. It can help relieve stress, for example, as well as combat the effects of aging, improve muscle coordination, relieve lower-back pain, and elongate muscles, making you look sleeker and fitter.

NECK SIDE BEND

① Stand tall, and gently grasp the side of your head with your hand.

BACK VIEW

② Reach toward the small of your back with your other hand, bending at the elbow.

ANNOTATION KEY

Black text indicates target muscles

Gray text indicates other working muscles

* indicates deep muscles

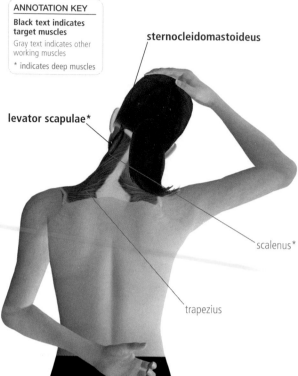

sternocleidomastoideus

levator scapulae*

scalenus*

trapezius

TARGETS
• Neck muscles

LEVEL
• Beginner

BENEFITS
• Increases neck flexibility

NOT ADVISABLE IF YOU HAVE . . .
• Severe neck pain

FRONT VIEW

③ Tilt your head toward your raised elbow until you feel the stretch in the side of your neck. Hold for 15 seconds, and repeat three times on each side.

AVOID
• Tensing or lifting your shoulders.
• Using your hand to tug your head downward

DO IT RIGHT
• Breathe easily and normally throughout the stretch.

BEST FOR

• levator scapulae
• sternocleidomastoideus

TRICEPS STRETCH

1. Stand tall keeping your neck, torso, and shoulders straight.

2. Raise your right arm, and bend it behind your head.

3. Keeping your shoulders relaxed, grasp your raised elbow with your left hand, and gently pull back.

4. Continue to pull your elbow back until you feel the stretch on the underside of your arm. Hold for 15 seconds.

5. Repeat three times on each arm.

BACK VIEW

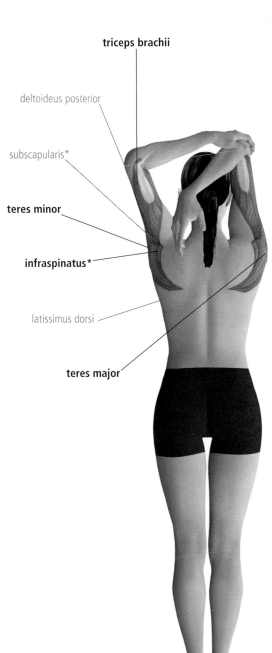

FRONT VIEW

BEST FOR

- triceps brachii
- infraspinatus
- teres major
- teres minor

DO IT RIGHT
- Keep your dropped elbow close to the side of your head.

AVOID
- Leaning backward.

triceps brachii

deltoideus posterior

subscapularis*

teres minor

infraspinatus*

latissimus dorsi

teres major

TARGETS
- Shoulders
- Triceps

LEVEL
- Beginner

BENEFITS
- Improves range of motion

NOT ADVISABLE IF YOU HAVE . . .
- Shoulder instability

ANNOTATION KEY
Black text indicates target muscles
Gray text indicates other working muscles
* indicates deep muscles

POSTERIOR HAND CLASP

❶ Stand tall, keeping your neck, shoulders, and torso straight. Your arms should hang loosely at your sides.

❷ Extend your right hand to the side, parallel to the floor.

TARGETS
• Upper back
• Upper arms

LEVEL
• Intermediate

BENEFITS
• Stretches the shoulders, chest, and upper arms

NOT ADVISABLE IF YOU HAVE . . .
• Shoulder injury

❸ Bend your elbow, and rotate your shoulder downward so that the palm of your hand faces outward. Reach behind your back, palm still up, and draw your elbow into your right side.

❹ Continue to rotate your shoulder downward as you reach upward with your hand until your forearm is parallel to your spine. Your right hand should rest in between your shoulder blades.

❺ Reach your left arm up with your palm facing directly behind you. Bend your elbow, reaching your left hand down the center of your back.

❻ Hook your hands together behind your back. Lift your chest, and pull your abdominals in toward your spine.

❼ Hold for about 30 seconds to 1 minute. Release your arms, and repeat with your arms reversed for the same length of time.

AVOID
• Straining—if you cannot hook your hands behind your back, try using a strap or an elastic exercise band to help you pull your hands closer together.

DO IT RIGHT
• Keep your lower elbow tucked close to the side of your torso.

BEST FOR

- rhomboideus
- teres minor
- subscapularis
- latissimus dorsi
- deltoideus anterior
- deltoideus medialis
- deltoideus posterior
- triceps brachii
- pectoralis major
- pectoralis minor

ANNOTATION KEY

Black text indicates target muscles
Gray text indicates other working muscles
* indicates deep muscles

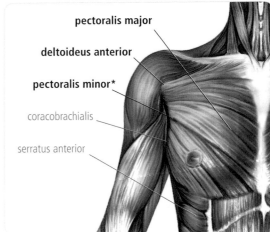

pectoralis major
deltoideus anterior
pectoralis minor*
coracobrachialis
serratus anterior

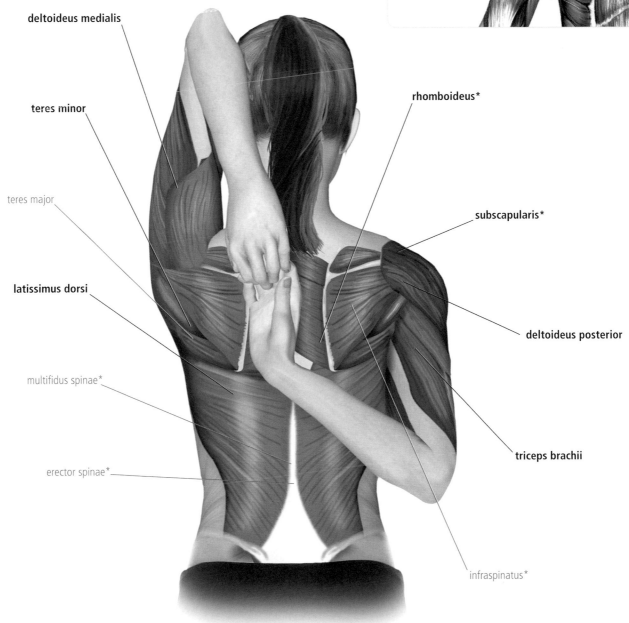

deltoideus medialis
teres minor
teres major
latissimus dorsi
multifidus spinae*
erector spinae*

rhomboideus*
subscapularis*
deltoideus posterior
triceps brachii
infraspinatus*

CHEST STRETCH

1 Stand with your hands behind your head, with fingers interlocked. Your elbows should be pointing outward.

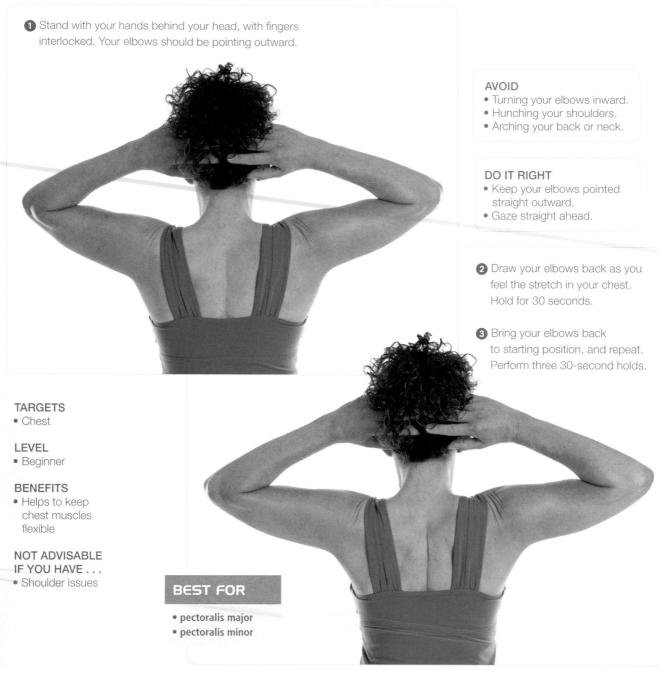

AVOID
• Turning your elbows inward.
• Hunching your shoulders.
• Arching your back or neck.

DO IT RIGHT
• Keep your elbows pointed straight outward.
• Gaze straight ahead.

2 Draw your elbows back as you feel the stretch in your chest. Hold for 30 seconds.

3 Bring your elbows back to starting position, and repeat. Perform three 30-second holds.

TARGETS
• Chest

LEVEL
• Beginner

BENEFITS
• Helps to keep chest muscles flexible

NOT ADVISABLE IF YOU HAVE . . .
• Shoulder issues

BEST FOR

• pectoralis major
• pectoralis minor

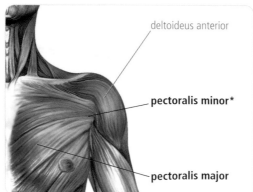

deltoideus anterior

pectoralis minor*

pectoralis major

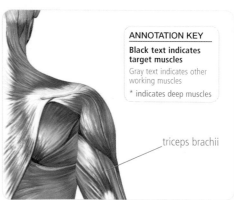

ANNOTATION KEY

Black text indicates target muscles
Gray text indicates other working muscles
* indicates deep muscles

triceps brachii

SWISS BALL KNEELING LAT STRETCH

1 Kneel on all fours in front of your Swiss ball. Extend one arm, placing your hand on the ball. Rest your other hand on the floor.

2 Lean back onto your heels until you feel a deep stretch in large muscles on either side of your back. Hold for 30 seconds.

3 Switch arms, and repeat. Complete three 30-second holds per arm.

BEST FOR
- **latissimus dorsi**
- **erector spinae**

TARGETS
- Back

LEVEL
- Beginner

BENEFITS
- Helps to keep back muscles flexible

AVOID IF YOU HAVE . . .
- Lower-back issues

infraspinatus*

supraspinatus*

deltoideus posterior

teres minor

subscapularis*

triceps brachii

latissimus dorsi

erector spinae*

AVOID
- Allowing your torso to twist.
- Arching your neck.

DO IT RIGHT
- Keep your arm fully extended on the ball.
- Face the floor throughout the stretch.

ANNOTATION KEY
Black text indicates target muscles

Gray text indicates other working muscles

* indicates deep muscles

21

LATISSIMUS DORSI STRETCH

1 Stand, keeping your neck, shoulders, and torso straight.

2 Raise both arms above your head and clasp your hands together, palms facing upward.

AVOID
• Leaning backward as you come to the top of the circle.

DO IT RIGHT
• Elongate your arms and shoulders as much as possible.

TARGETS
• Back
• Obliques

LEVEL
• Beginner

BENEFITS
• Helps correct poor posture

NOT ADVISABLE IF YOU HAVE . . .
• Lower back pain

3 Keeping your elbows straight, reach to the side to begin tracing a circular pattern with your torso.

BEST FOR

• latissimus dorsi
• obliquus externus

4 Lean forward and then to the opposite side as you slowly trace a full circle.

5 Return to the starting position, and then repeat the sequence three times in each direction.

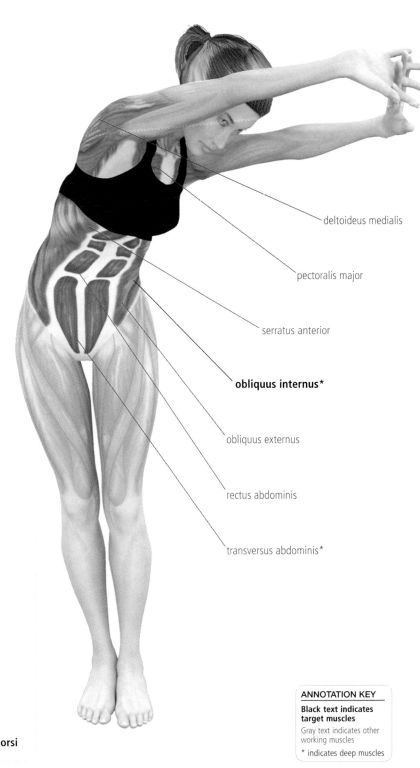

deltoideus medialis

pectoralis major

serratus anterior

obliquus internus*

obliquus externus

rectus abdominis

transversus abdominis*

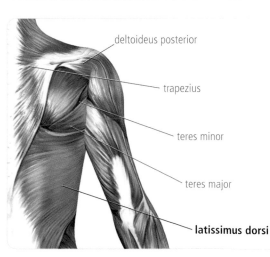

deltoideus posterior

trapezius

teres minor

teres major

latissimus dorsi

ANNOTATION KEY
Black text indicates target muscles
Gray text indicates other working muscles
* indicates deep muscles

TOE TOUCH

❶ Stand up tall, and exhale.

DO IT RIGHT
- Stack your spine one vertebra at a time.
- Connect the stretch in your back with the stretch in your hamstrings.
- Make the stretch long and smooth.

AVOID
- Tensing your neck muscles.
- Bouncing as you try to reach your hands to your toes—reach down only as far as you can comfortably extend.

TARGETS
- Spine

LEVEL
- Beginner

BENEFITS
- Stretches the spine and hamstrings
- Refines spinal stacking skills

NOT ADVISABLE IF YOU HAVE . . .
- Lower-back pain that radiates down the leg

❷ Tucking your head down toward your chest and rolling down one vertebra at a time, reach down toward your toes. Keeping your weight slightly shifted forward, continue exhaling, rounding your spine.

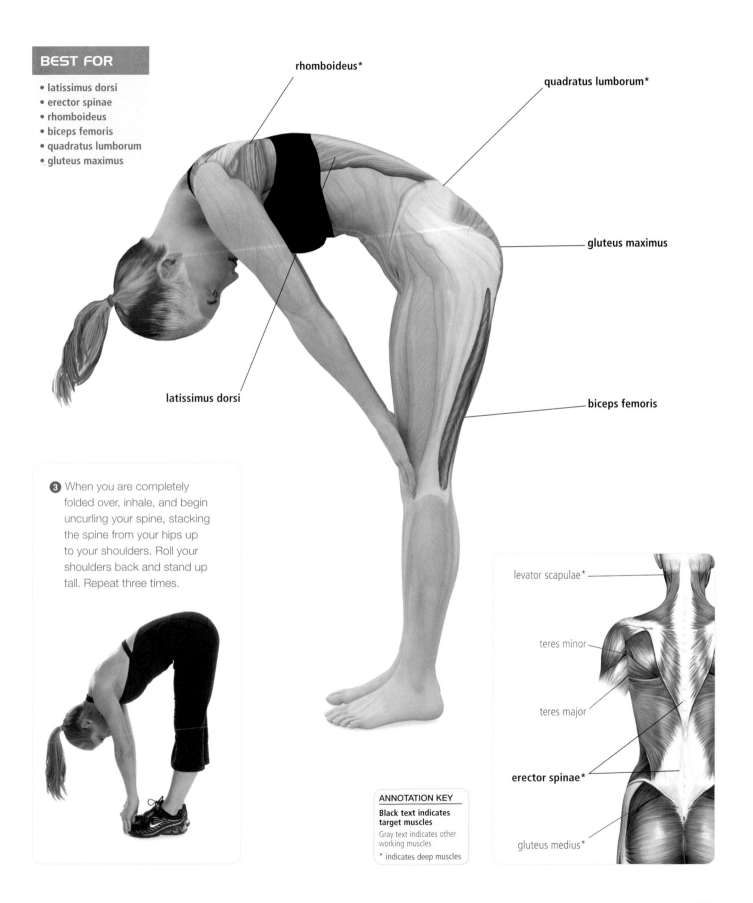

BEST FOR

- latissimus dorsi
- erector spinae
- rhomboideus
- biceps femoris
- quadratus lumborum
- gluteus maximus

rhomboideus*

quadratus lumborum*

gluteus maximus

latissimus dorsi

biceps femoris

3 When you are completely folded over, inhale, and begin uncurling your spine, stacking the spine from your hips up to your shoulders. Roll your shoulders back and stand up tall. Repeat three times.

levator scapulae*

teres minor

teres major

erector spinae*

gluteus medius*

ANNOTATION KEY

Black text indicates target muscles

Gray text indicates other working muscles

* indicates deep muscles

CAT AND DOG STRETCH

❶ Kneel on all fours, with your wrists directly below your shoulders and your knees directly below your hips. Your fingertips should be facing forward, with your hands shoulder-width apart. Look down at the floor, keeping your head in a neutral position.

AVOID
- Arching primarily in your lower back.
- Tucking your chin to your chest in the Cat phase of the stretch.
- Jutting out your rib cage in the Dog phase of the stretch.

❷ Exhale, and round your spine up toward the ceiling, dropping your head. Draw your belly button in toward your spine. Keep your hips lifted and your shoulders in the same position. This is the Cat phase of the stretch.

TARGETS
- Lower- and middle-back extensors
- Abdominals
- Obliques

LEVEL
- Beginner

BENEFITS
- Stretches chest, shoulders, neck, spine, and abdominals
- Improves range of motion

NOT ADVISABLE IF YOU HAVE . . .
- Knee injury
- Wrist pain

❸ Inhale, and uncurl your spine. Remain on your hands and knees.

❹ With your next inhalation, arch your spine, lifting your chest forward and your tailbone toward the ceiling. Look forward. This is the Dog phase of the stretch.

❺ Exhale, and return to a neutral position on your hands and knees.

❻ Repeat the entire sequence 10 to 20 times.

BEST FOR

- **erector spinae**

ANNOTATION KEY

Bold text indicates target muscles

Gray text indicates other working muscles

Italics indicates ligaments

* indicates deep muscles

DO IT RIGHT

- Stretch slowly and with control.
- Keep your hands and feet planted throughout the stretch.
- Lift your chin while your spine is arched.
- Start the movement of your spine in your tailbone.

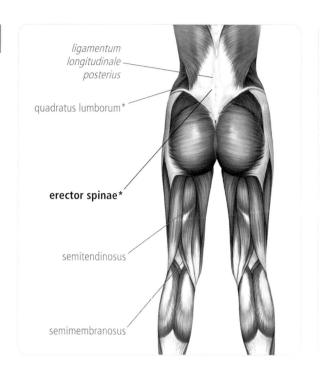

ligamentum longitudinale posterius

quadratus lumborum*

erector spinae*

semitendinosus

semimembranosus

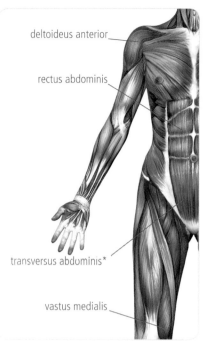

deltoideus anterior

rectus abdominis

transversus abdominis*

vastus medialis

latissimus dorsi

gluteus medius*

rhomboideus*

gluteus maximus

deltoideus posterior

deltoideus medialis

biceps femoris

PIRIFORMIS STRETCH

❶ Lie on your back with your knees bent.

❷ Bring your left ankle over your right knee, resting it on your thigh. Place both hands around your right thigh.

❸ Gently pull your right thigh toward your chest until you feel the stretch in your buttocks. Hold for 15 seconds and switch sides. Repeat sequence on your left leg.

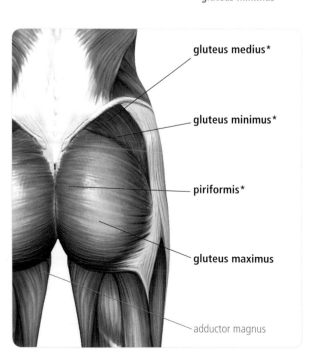

TARGETS
• Gluteal muscles

LEVEL
• Beginner

BENEFITS
• Stretches the glutes

NOT ADVISABLE IF YOU HAVE . . .
• Hip dysfunction

DO IT RIGHT
• Relax your hips so that you can go deeper into the stretch.
• Perform the stretch slowly
• Keep your head and shoulders on the floor

AVOID
• Pulling your leg inward too quickly.
• Twisting your lower body—instead keep your hips square

ANNOTATION KEY
Black text indicates target muscles
Gray text indicates other working muscles
* indicates deep muscles

BEST FOR
• **piriformis**
• **gluteus maximus**
• **gluteus medius**
• **gluteus minimus**

gluteus medius*

gluteus minimus*

piriformis*

gluteus maximus

adductor magnus

HIP STRETCH

BEST FOR

- adductor longus
- iliopsoas
- rhomboideus
- sternocleidomastoideus
- latissimus dorsi
- obliquus internus
- obliquus externus
- quadratus lumborum
- erector spinae
- multifidus spinae
- tractus iliotibialis
- gluteus maximus
- gluteus medius
- piriformis

DO IT RIGHT
- Keep your neck and shoulders relaxed.
- Apply even pressure to your leg with your active hand.
- Keep torso upright as you pull your knee and torso together.

AVOID
- Rounding your torso.
- Lifting the foot of your bent leg off the floor.
- Straining your neck as you rotate.

① Sit with your left leg extended straight in front of you, and bend your right knee. Cross your bent knee over the straight leg, and keep your foot flat on the floor.

② Wrap your left arm around the bent knee so that you are able to apply pressure to your leg to rotate your torso. Place your right hand on the floor for stability.

③ Keeping your hips aligned, rotate your upper spine as you pull your chest in toward your knee.

④ Hold for 30 seconds. Slowly release, and repeat five times on each side.

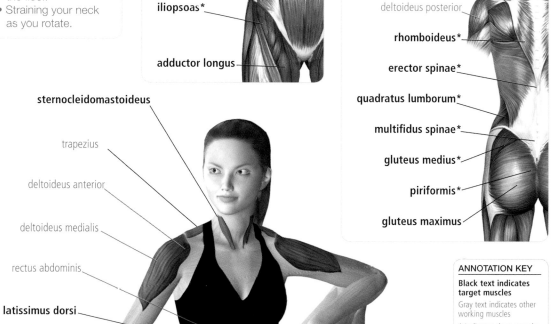

iliopsoas*

adductor longus

sternocleidomastoideus

trapezius

deltoideus anterior

deltoideus medialis

rectus abdominis

latissimus dorsi

obliquus externus

obliquus internus*

tractus iliotibialis

adductor magnus

deltoideus posterior

rhomboideus*

erector spinae*

quadratus lumborum*

multifidus spinae*

gluteus medius*

piriformis*

gluteus maximus

TARGETS
- Hips
- Gluteal muscles
- Spine
- Obliques

LEVEL
- Intermediate

BENEFITS
- Stretches hip extensors and flexors
- Stretches obliques

NOT ADVISABLE IF YOU HAVE . . .
- Hip dysfunction
- Severe lower-back pain

ANNOTATION KEY
Black text indicates target muscles
Gray text indicates other working muscles
* indicates deep muscles

HIP-TO-THIGH STRETCH

❶ Kneeling on your left knee, place your right foot on the floor in front of you so that your right knee is bent less than 90 degrees.

❷ Bring your torso forward, bending your right knee so that your knee shifts toward your toes. Keeping your torso in neutral position, press your right hip forward and downward to create a stretch over the front of your thigh. Raise your arms up toward the ceiling, keeping your shoulders relaxed.

❸ Bring your arms down and move your hips backward. Straighten your right leg, and bring your torso forward. Place your hands on either side of your straight leg for support.

❹ Hold for 10 seconds, and repeat the forward and backward movement five times on each leg.

DO IT RIGHT
• Keep your shoulders and neck relaxed.
• Move your entire body as one unit as you go into the stretch.

AVOID
• Extending your front knee too far over the planted foot.
• Rotating your hips.
• Shifting your back knee outward.

TARGETS
• Hip flexors
• Hip extensors
• Hamstrings
• Quadriceps

LEVEL
• Intermediate

BENEFITS
• Stretches hips and thighs

NOT ADVISABLE IF YOU HAVE . . .
• Neck pain
• Lower-back pain

MODIFICATION
Harder: During the backward movement, raise your back knee off the floor and straighten your back leg. Keep your hands on the floor.

BEST FOR
• iliopsoas
• biceps femoris
• rectus femoris

Black text indicates target muscles
Gray text indicates other working muscles
* indicates deep muscles

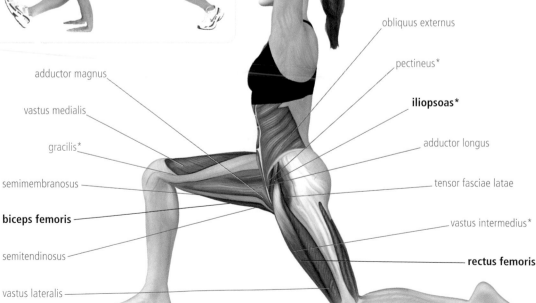

adductor magnus

vastus medialis

gracilis*

semimembranosus

biceps femoris

semitendinosus

vastus lateralis

obliquus externus

pectineus*

iliopsoas*

adductor longus

tensor fasciae latae

vastus intermedius*

rectus femoris

SPINE STRETCH

1 Lie on your back with your left leg straight and the right leg bent, placing your right foot on your left shin.

BEST FOR

- quadratus lumborum
- erector spinae
- tractus iliotibialis
- tensor fasciae latae

2 Keeping both shoulders on the floor, slowly bring your right leg across your body until you feel the stretch in the area between your lower back and hips. Stretch only as far as your shoulders will allow without one of them rising from the floor.

3 Hold for 15 seconds, and repeat sequence three times on each side.

TARGETS
- Spinal extensors

LEVEL
- Beginner

BENEFITS
- Stretches lower back

NOT ADVISABLE IF YOU HAVE . . .
- Hip issues

ANNOTATION KEY

Black text indicates target muscles

Gray text indicates other working muscles

* indicates deep muscles

AVOID
- Allowing your shoulders to lift off the floor.

DO IT RIGHT
- Keep your lower back relaxed.

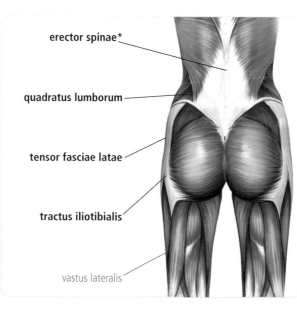

erector spinae*

quadratus lumborum

tensor fasciae latae

tractus iliotibialis

vastus lateralis

SWISS BALL HIP CROSSOVER

1 Lie on your back, with your arms extended out to your sides. Place your legs on a Swiss ball, with glutes close to the ball.

2 Brace your abs, and lower your legs to one side, as close to the floor as you can possibly go without raising your shoulders off the floor.

TARGETS
• Lower back
• Obliques

LEVEL
• Intermediate

BENEFITS
• Helps to strengthen and tone abs
• Improves core stabilization

NOT ADVISABLE IF YOU HAVE . . .
• Lower-back issues

AVOID
• Swinging your legs too quickly; instead, try to keep the movement smooth and controlled.

3 Return to the starting position, and then repeat on the other side. Work up to completing 20 in each direction.

MODIFICATION

Easier: Begin with your legs lifted and bent at a 90-degree angle. Try to keep your upper body as stable as possible as you perform the crossover without the ball, alternating sides.

BEST FOR

- erector spinae
- obliquus externus

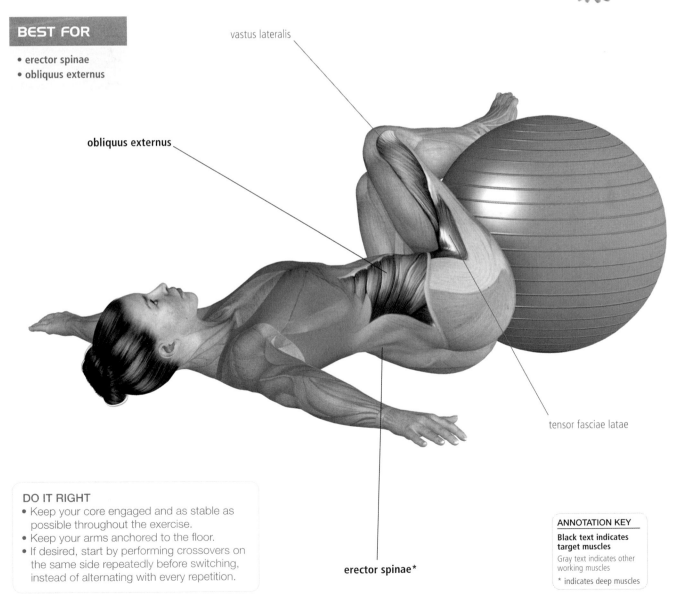

vastus lateralis

obliquus externus

tensor fasciae latae

erector spinae*

DO IT RIGHT

- Keep your core engaged and as stable as possible throughout the exercise.
- Keep your arms anchored to the floor.
- If desired, start by performing crossovers on the same side repeatedly before switching, instead of alternating with every repetition.

ANNOTATION KEY

Black text indicates target muscles

Gray text indicates other working muscles

* indicates deep muscles

KNEE-TO-CHEST HUG

❶ Lie supine on a mat with your legs together and arms outstretched.

❷ Bend your right knee, and bring your foot to your body's midline while clasping your hands together to hold your knee. Hold the stretch for 15 seconds.

TARGETS
• Lower back
• Hips

LEVEL
• Beginner

BENEFITS
• Stretches lower back, hip extensors, and hip rotators

NOT ADVISABLE IF YOU HAVE . . .
• Advanced degenerative joint disease

❸ Return to the starting position.

❹ Again, clasping your hands together to hold your knee, bend your right knee, but this time rotate the right leg to the left, bringing the side of your leg against your chest.

❺ Hold the stretch for 15 seconds, and then return to the starting position. Repeat the entire sequence with the left leg bent.

MODIFICATION

Similar level of difficulty: Follow step 1, and then draw both legs to your chest.

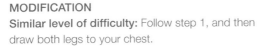

BEST FOR

- erector spinae
- latissimus dorsi
- gluteus maximus
- gluteus minimus
- piriformis
- gemellus superior
- gemellus inferior
- obturator externus
- obturator internus
- quadratus femoris

AVOID

- Lifting your buttocks off the floor.

DO IT RIGHT

- Keep your spine in neutral position.

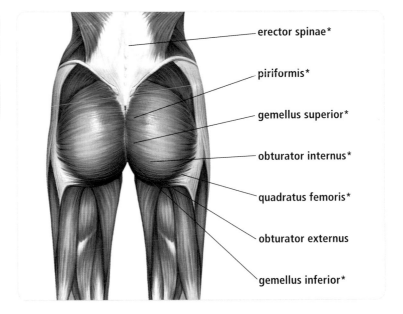

erector spinae*

piriformis*

gemellus superior*

obturator internus*

quadratus femoris*

obturator externus

gemellus inferior*

ANNOTATION KEY

Black text indicates target muscles

Gray text indicates other working muscles

* indicates deep muscles

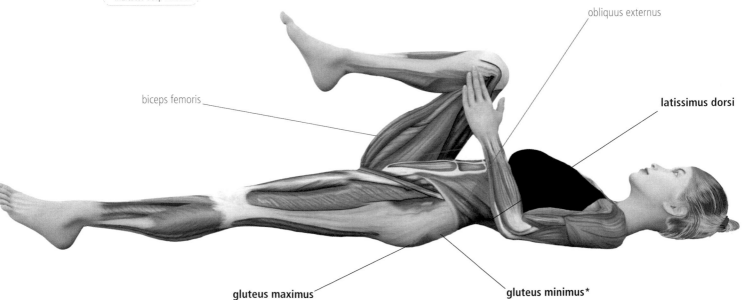

obliquus externus

biceps femoris

latissimus dorsi

gluteus maximus

gluteus minimus*

ILIOTIBIAL BAND STRETCH

1 Standing, cross your left leg in front of your right.

2 Bend forward from the hips while keeping both legs straight, and reach your hands toward the floor.

3 Hold for 15 seconds. Repeat the sequence three times on each leg.

BEST FOR

- **tractus iliotibialis**
- **biceps femoris**
- **gluteus maximus**
- **vastus lateralis**

TARGETS
- Iliotibial band
- Hamstrings

LEVEL
- Beginner

BENEFITS
- Helps to stabilize knee joints
- Helps to keep hips flexible
- Stretches back, hamstrings, and calves

NOT ADVISABLE IF YOU HAVE . . .
- Neck issues
- Lower-back pain

AVOID
- Raising your back heel off the floor.
- Arching or round your back.

DO IT RIGHT
- Keep both feet flat on the floor.
- Stretch with good alignment so that your back leg and your spine form a straight line

tractus iliotibialis

gluteus maximus

biceps femoris

rectus femoris

semitendinosus

semimembranosus

vastus lateralis

gastrocnemius

soleus

ANNOTATION KEY
Black text indicates target muscles
Gray text indicates other working muscles

QUADRICEPS STRETCH

BEST FOR

- rectus femoris
- vastus lateralis
- vastus medialis
- vastus intermedius

1 Stand with your feet together. Bend your left leg behind you, and grasp your foot with your left hand. Pull your heel toward your buttocks until you feel a stretch in the front of your thigh. Keep both knees together and aligned.

2 Hold for 15 seconds. Repeat sequence three times on each leg.

tensor fasciae latae

vastus intermedius*

rectus femoris

vastus lateralis

vastus medialis

TARGETS
- Quadriceps

LEVEL
- Beginner

BENEFITS
- Helps to keep thigh muscles flexible

NOT ADVISABLE IF YOU HAVE . . .
- Knee issues

DO IT RIGHT
- Both knees to remain pressed together.

AVOID
- Leaning forward with your chest.

ANNOTATION KEY

Black text indicates target muscles

Gray text indicates other working muscles

STANDING HAMSTRINGS STRETCH

❶ Stand with one leg bent and the other extended in front of you with the heel on the floor.

FRONT VIEW

BEST FOR

- biceps femoris
- semitendinosus
- semimembranosus

BACK VIEW

DO IT RIGHT
- Keep your front leg straight.
- Flex the foot of your front leg as you stretch.

TARGETS
- Hamstrings

LEVEL
- Beginner

BENEFITS
- Helps to keep hamstring muscles flexible

NOT ADVISABLE IF YOU HAVE . . .
- Lower-back issues
- Knee issues

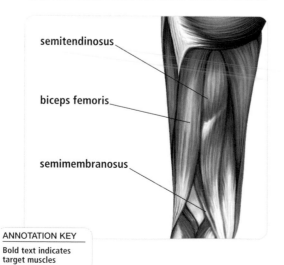

semitendinosus

biceps femoris

semimembranosus

ANNOTATION KEY
Bold text indicates
target muscles

❷ Lean over your extended leg, resting both hands above your knee. Place the majority of your body weight on your front heel while feeling the stretch in the back of your thigh. Hold for 30 seconds.

❸ Switch sides and repeat. Complete three 30-second holds on each leg.

AVOID
- Allowing your back to round forward.
- Hunching your shoulders.

STANDING CALF STRETCH

1. Stand with one foot in front of the other, with the front leg bent. With a straight back, lean over your front leg, resting both hands above the knee.

2. Place the majority of your body weight on your front heel as you feel the stretch in the calf muscle of your back leg. Hold for 30 seconds.

3. Switch sides and repeat. Complete three 30-second holds per leg.

BEST FOR

• **gastrocnemius**

TARGETS
• Calves

LEVEL
• Intermediate

BENEFITS
• Helps to keep calf muscles flexible

AVOID IF YOU HAVE . . .
• Knee issues

DO IT RIGHT
• Keep both feet flat on the floor.
• Stretch with good alignment so that your back leg and your spine form a straight line.

AVOID
• Raising your back heel off the floor.
• Arching or round your back.

ANNOTATION KEY

Black text indicates target muscles

Gray text indicates other working muscles

Italics indicates tendons

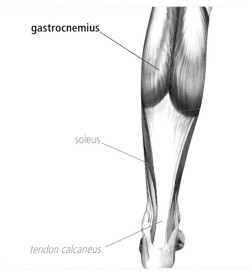

gastrocnemius

soleus

tendon calcaneus

CHILD'S POSE

❶ Kneel on a mat with your hips aligned over your knees. Bring your legs together so that your big toes are touching.

❷ Sit back, resting your buttocks on your heels. Separate your knees about hip-width apart.

DO IT RIGHT
• Round your back to create a dome shape.

BEST FOR

• latissimus dorsi
• trapezius
• deltoideus anterior
• deltoideus posterior
• rhomboideus
• teres major
• serratus anterior
• gluteus maximus
• erector spinae
• quadratus lumborum

TARGETS
• Lower back

LEVEL
• Beginner

BENEFITS
• Stretches and relaxes the back

NOT ADVISABLE IF YOU HAVE . . .
• Knee injury

❸ Lower your chest onto your thighs as you extend your hands in front of your head, elongating your neck and spine as you stretch your tailbone toward the mat.

❹ Place your forehead on the mat, and hold this position for 30 seconds to 3 minutes.

40

AVOID
- Rushing through the exercise. It can take a few minutes to allow your body to deepen into the full stretch.
- Compressing the back of your neck.

splenius*

deltoideus posterior

teres minor

teres major

erector spinae*

quadratus lumborum*

trapezius

rhomboideus*

latissimus dorsi

deltoideus anterior

serratus anterior

ANNOTATION KEY

Black text indicates target muscles

Gray text indicates other working muscles

* indicates deep muscles

brachialis

gluteus maximus

biceps brachii

vastus lateralis

extensor carpi radialis

triceps brachii

flexor digitorum*

UPPER-BODY EXERCISES

What woman doesn't want great looking arms? Toned arms make you look younger and stronger, and allow you to move freely without worrying about any embarrassing jiggling. For sleek, shapely arms that you'll want to show off in tank tops and strapless dresses focus on exercises that work your deltoids, biceps, and triceps. Also included in this section are exercises that target other upper-body muscles, especially the upper back and chest. Upper-back exercises define and enhance the shape of your neck and shoulders, and reduce any unsightly back fat—those bulges above and below your bra strap. Exercises that tone the chest muscles (the major and minor pectorals) have another major perk: they work as natural breast lifts, giving you a more youthful, fitter silhouette.

CHAIR DIP

1. Sit up tall near the front of a sturdy chair. Place your hands beside your hips, wrapping your fingers over the front edge of the chair.

2. Extend your legs in front of you slightly, and place your feet flat on the floor.

3. Scoot off the edge of the chair until your knees align directly above your feet and your torso will be able to clear the chair as you dip down.

DO IT RIGHT
- Keep your body close to the chair.
- Keep your spine in neutral position throughout the movement.

TARGETS
- Triceps
- Shoulder and core stabilizers

LEVEL
- Intermediate

BENEFITS
- Strengthens the shoulder girdle
- Trains the torso to remain stable while the legs and arms are in motion

NOT ADVISABLE IF YOU HAVE . . .
- Shoulder pain
- Wrist pain

4. Bending your elbows directly behind you, without splaying them out to the sides, lower your torso until your elbows make a 90-degree angle.

5. Press into the chair, raising your body back to the starting position. Repeat 15 times for two sets.

BEST FOR

- deltoideus posterior
- triceps brachii
- pectoralis major
- pectoralis minor
- latissimus dorsi
- rectus abdominis

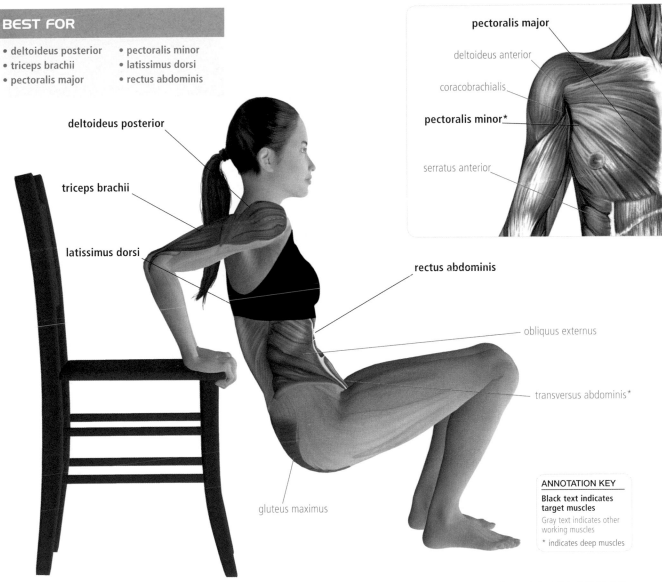

deltoideus posterior

triceps brachii

latissimus dorsi

pectoralis major

deltoideus anterior

coracobrachialis

pectoralis minor*

serratus anterior

rectus abdominis

obliquus externus

transversus abdominis*

gluteus maximus

ANNOTATION KEY

Black text indicates target muscles

Gray text indicates other working muscles

* indicates deep muscles

MODIFICATION

Harder: Keeping your knees squeezed together, perform the dips with one leg lifted straight out, parallel to the floor. Repeat 15 times on each side.

AVOID

- Allowing your shoulders to lift toward your ears.
- Moving your feet.
- Rounding your back at your hips.
- Pushing up solely with your feet, rather than using your arm strength.

CHAIR CRUNCH

❶ Sit up tall on a chair with your hands grasping the sides of the seat and your arms straight.

❷ Step forward so that your knees are bent but your buttocks are lifted off the chair. Your hips and knees should be bent to form 90-degree angles.

DO IT RIGHT
- Keep your spine in neutral position as you progress through the motion.
- Align your knees over your ankles.
- Keep your body close to the chair.

AVOID
- Allowing your shoulders to lift toward your ears.

TARGETS
- Shoulders
- Upper arms
- Abdominals

LEVEL
- Advanced

BENEFITS
- Strengthens upper body
- Improves shoulder stability

NOT ADVISABLE IF YOU HAVE . . .
- Shoulder pain
- Neck pain

❸ Tuck your tailbone toward the front of the chair, and bend your knees toward your chest. Bend your elbows simultaneously. At the bottom of the movement, extend your elbows and press through your shoulders.

❹ Keeping your head in neutral position, press into the chair and lower your legs to return to the starting position. Repeat 15 times for two sets.

BEST FOR

- triceps brachii
- deltoideus anterior
- deltoideus medialis
- deltoideus posterior
- infraspinatus
- supraspinatus
- teres minor
- subscapularis
- iliopsoas
- gracilis
- rectus abdominis
- transversus abdominis

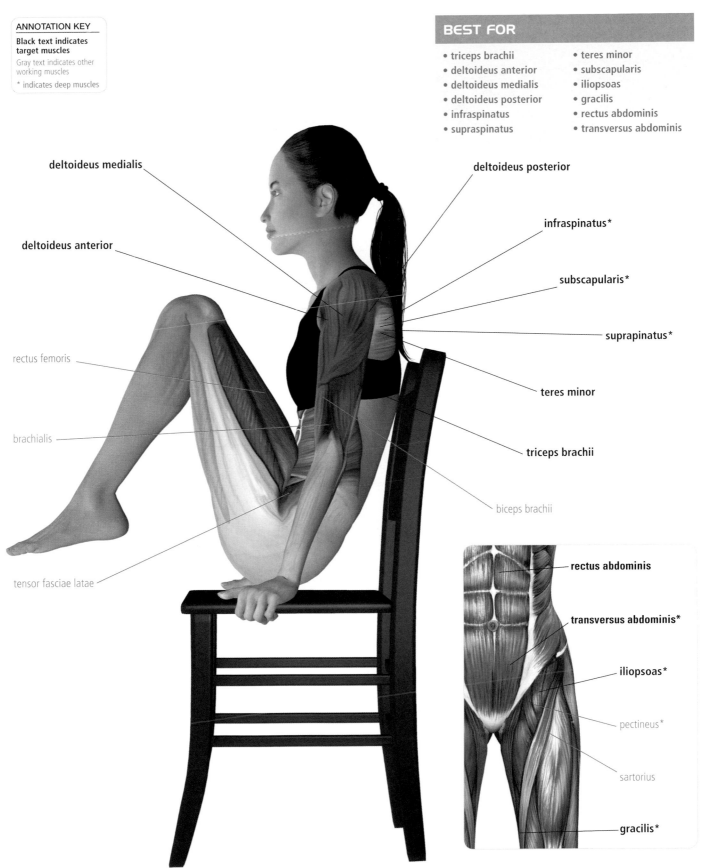

deltoideus medialis

deltoideus anterior

rectus femoris

brachialis

tensor fasciae latae

deltoideus posterior

infraspinatus*

subscapularis*

suprapinatus*

teres minor

triceps brachii

biceps brachii

rectus abdominis

transversus abdominis*

iliopsoas*

pectineus*

sartorius

gracilis*

OVERHEAD PRESS

1 Stand upright with one leg extended about a foot behind you, heel off the ground. Position a resistance band beneath the foot of your front leg. Hold the handles in both hands, with arms bent, so that the resistance band is taut.

2 Straighten both arms so that they are extended to full lockout above your head a few inches in front of your shoulders.

TARGETS
• Shoulders
• Triceps

LEVEL
• Beginner

BENEFITS
• Strengthens and tones shoulders and upper arms

NOT ADVISABLE IF YOU HAVE . . .
• Shoulder issues

3 Lower your arms to starting position and then repeat. Perform three sets of 15.

DO IT RIGHT
• Keep the rest of your body stable as you extend your arms.
• Gaze forward throughout the exercise.
• Keep your abs engaged and pulled in.
• Extend both arms at the same time.

AVOID
• Twisting your torso.

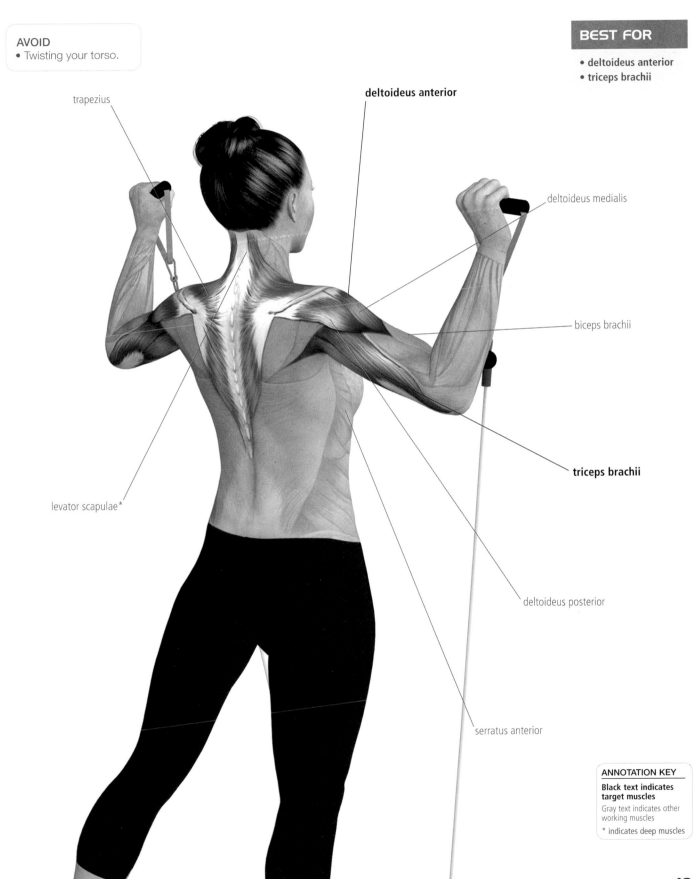

trapezius

deltoideus anterior

deltoideus medialis

biceps brachii

triceps brachii

levator scapulae*

deltoideus posterior

serratus anterior

ANNOTATION KEY

Black text indicates target muscles

Gray text indicates other working muscles

* indicates deep muscles

ALTERNATING CHEST PRESS

❶ Run a resistance band around a sturdy, stable object, such as a pole or column. Stand facing away from the object, holding both ends of the resistance band in front of your chest.

AVOID
- Twisting your torso.
- Hunching your shoulders.

❷ Extend one arm straight in front of you to full lockout position, keeping the other arm steady.

❸ With control, bring the arm back to starting position. Repeat with the other arm, completing three sets of 15 repetitions per arm.

TARGETS
- Chest
- Core
- Shoulders
- Triceps

LEVEL
- Beginner

BENEFITS
- Strengthens and tones pectorals
- Stabilizes core

NOT ADVISABLE IF YOU HAVE . . .
- Shoulder issues

DO IT RIGHT
- Keep one arm motionless as you extend the other to lockout.
- Maintain a stable torso.
- Keep your feet in place as you extend your arm.
- Engage your abs throughout the exercise.
- Keep your arms level with your shoulders.

ANNOTATION KEY

Black text indicates target muscles

Gray text indicates other working muscles

* indicates deep muscles

BEST FOR

• pectoralis major

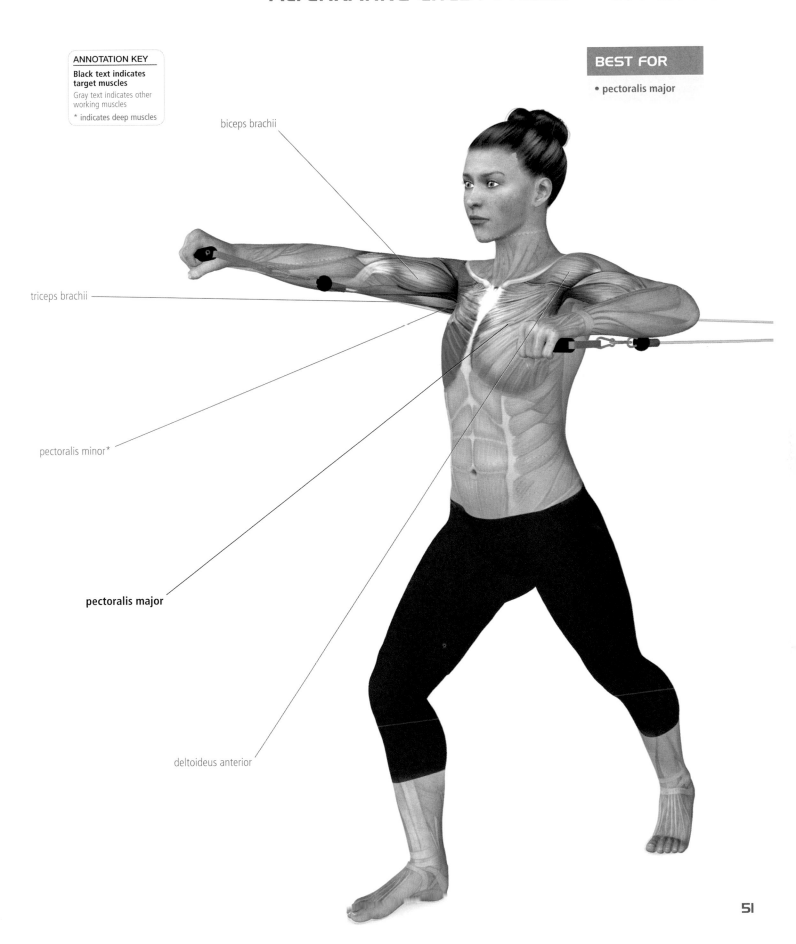

biceps brachii

triceps brachii

pectoralis minor*

pectoralis major

deltoideus anterior

STANDING FLY

❶ Run a resistance band around a sturdy, stable object such as a pole or column. Stand upright, with your feet planted shoulder-width apart and your knees soft. Grasp both of the handles of your resistance band, and extend your arms in front of you to almost shoulder height, holding the band taut.

BEST FOR

• pectoralis major

TARGETS
• Back
• Chest
• Upper back

LEVEL
• Beginner

BENEFITS
• Strengthens and tones shoulders and upper back

NOT ADVISABLE IF YOU HAVE . . .
• Lower-back issues
• Shoulder pain

❷ Slowly and with control, bring both arms out to the sides.

❸ Return to starting position and repeat. Complete three sets of 15 repetitions.

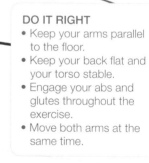

DO IT RIGHT
• Keep your arms parallel to the floor.
• Keep your back flat and your torso stable.
• Engage your abs and glutes throughout the exercise.
• Move both arms at the same time.

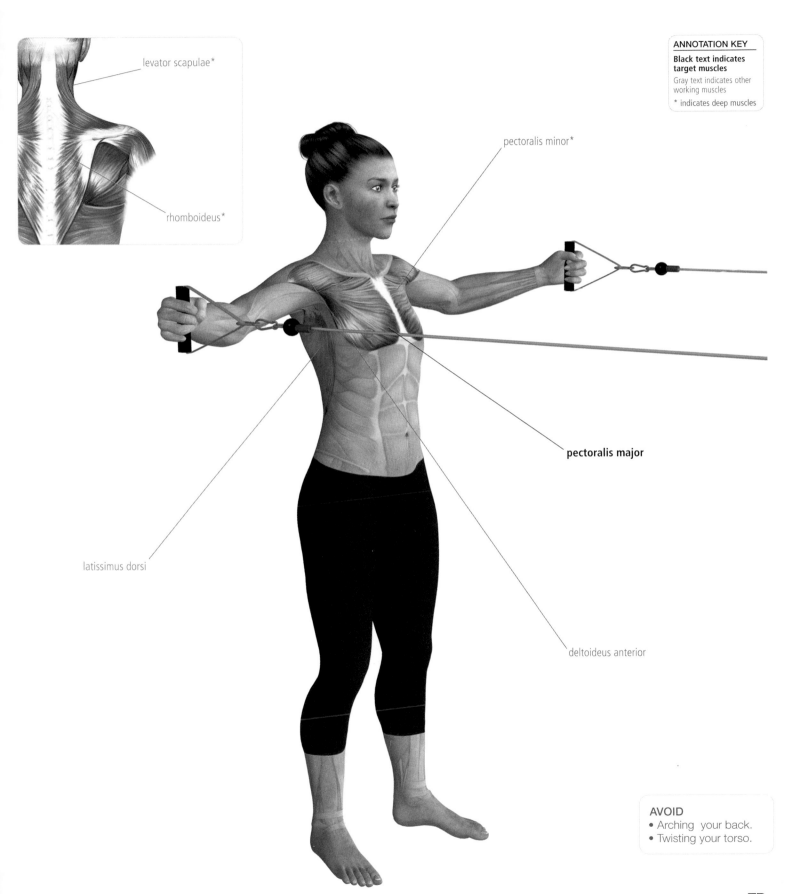

levator scapulae*

rhomboideus*

pectoralis minor*

pectoralis major

latissimus dorsi

deltoideus anterior

AVOID
• Arching your back.
• Twisting your torso.

UPWARD PLANK

1 Sit with your legs extended, and place the palms of your hands on the floor.

DO IT RIGHT
- Use your hamstrings and shoulders to open your hips and chest, rather than overextend your back. If your hamstrings are too weak, keep your legs bent while holding the lift in your hips.
- Breathe steadily, using the breath to deepen the extension in your upper back.

2 Move your hands so that your palms rest several inches behind your hips, fingers facing forward.

3 Draw your knees toward your chest. Place your feet on the floor with your heels about 12 inches away from your buttocks, and turn your big toes slightly inward.

TARGETS
- Upper back
- Upper arms
- Shoulders
- Chest
- Hamstrings

LEVEL
- Intermediate

BENEFITS
- Strengthens the shoulders, spine, arms, and hamstrings
- Stretches the hips and chest

NOT ADVISABLE IF YOU HAVE . . .
- Neck injury
- Wrist injury

4 Exhale, pressing down with your hands and feet and lifting your hips until your back and thighs are parallel to the floor. Your shoulders should be directly above your wrists.

5 Without lowering your hips, straighten your legs one at a time.

6 Lifting your chest and bringing your shoulder blades together, push your hips higher, creating a slight arch in your back.

6 Gently elongate your neck, and let it drop back.

7 Hold for 30 seconds, and return to a seated position.

BEST FOR

- deltoideus medialis
- deltoideus anterior
- deltoideus posterior
- triceps brachii
- teres major
- teres minor
- erector spinae
- gluteus maximus
- gluteus medius
- adductor magnus
- biceps femoris
- pectoralis major

AVOID
- Squeezing your glutes to create the lift.
- Using your glutes to maintain the position.
- Allowing your hips to sag.

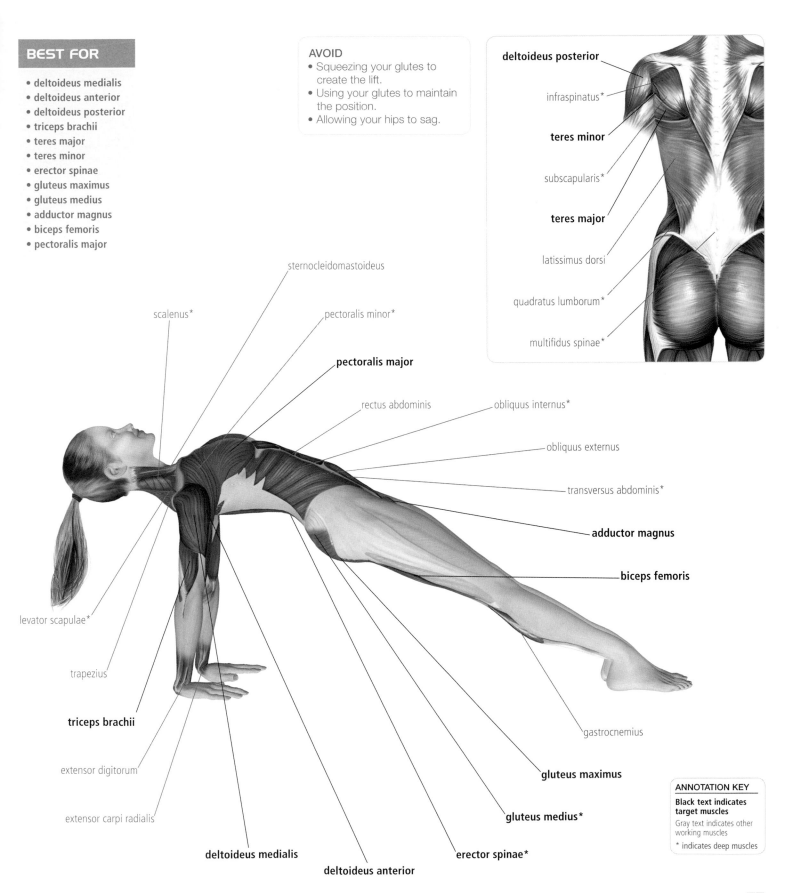

deltoideus posterior

infraspinatus*

teres minor

subscapularis*

teres major

latissimus dorsi

quadratus lumborum*

multifidus spinae*

sternocleidomastoideus

scalenus*

pectoralis minor*

pectoralis major

rectus abdominis

obliquus internus*

obliquus externus

transversus abdominis*

adductor magnus

biceps femoris

levator scapulae*

trapezius

triceps brachii

extensor digitorum

extensor carpi radialis

gastrocnemius

gluteus maximus

gluteus medius*

erector spinae*

deltoideus medialis

deltoideus anterior

ANNOTATION KEY

Black text indicates target muscles

Gray text indicates other working muscles

* indicates deep muscles

SWISS BALL PULLOVER

❶ Lie face-up on a Swiss ball, with your upper back, neck, and head supported. Your body should be extended with your torso long, knees bent at a right angle and feet planted on the floor a little wider than shoulder-distance apart. Grasp a hand weight or dumbbell in each hand, and extend your arms behind you, level with your shoulders so that your body from knees to fingertips forms a straight line.

AVOID
• Locking your arms when they are extended behind your head.
• Arching your back.
• Rushing through the exercise.

BEST FOR

• latissimus dorsi

TARGETS
• Upper back
• Core

LEVEL
• Intermediate

BENEFITS
• Strengthens upper back
• Stabilizes core

NOT ADVISABLE IF YOU HAVE . . .
• Shoulder issues

❷ Keeping the rest of your body stable and your arms as straight as possible, raise your arms upward so that they are perpendicular to your body.

❸ Return your arms to starting position. Repeat, performing three sets of 15 repetitions.

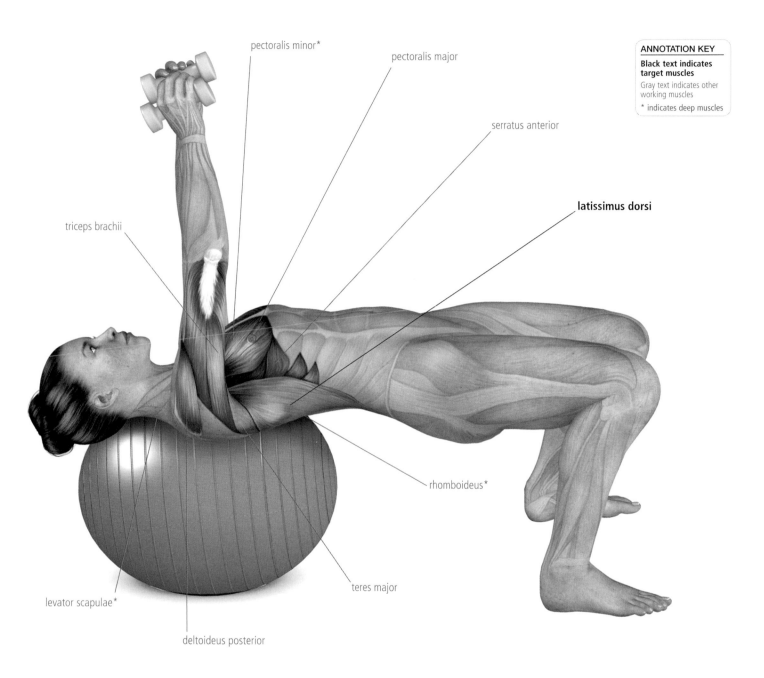

pectoralis minor*

pectoralis major

serratus anterior

latissimus dorsi

triceps brachii

rhomboideus*

levator scapulae*

teres major

deltoideus posterior

DO IT RIGHT
- Ease into the movement.
- Keep your arms directly above your shoulders when lifting the weights overhead.
- Keep your torso stable and feet planted throughout the exercise.
- Engage your abs.
- Keep your buttocks and pelvis lifted so that your upper legs, torso, and neck form a straight line.
- Move your arms smoothly and with control.

MODIFICATION
Similar level of difficulty: Instead of using hand weights, grasp a medicine ball in your hands as you perform the exercise.

SWISS BALL TRICEPS EXTENSION

1 Lie face-up on a Swiss ball, with your upper back, neck, and head supported. Your body should be extended with your torso long, knees bent at a right angle and feet planted on the floor a little wider than shoulder-distance apart. Grasp a hand weight or dumbbell in each hand and extend your arms straight up.

DO IT RIGHT
- Keep your forearms stable and your elbows over your shoulders.
- Keep your torso stable and feet planted throughout the exercise.
- Engage your abs.
- Keep your glutes and pelvis lifted so that your upper legs, torso, and neck form a straight line.
- Move smoothly and with control.

BEST FOR
- triceps brachii

TARGETS
- Triceps

LEVEL
- Intermediate

BENEFITS
- Strengthens and tones triceps

NOT ADVISABLE IF YOU HAVE . . .
- Elbow pain

2 Bend your elbows as you lower the weights toward your head.

3 Straighten your arms upward to starting position and then repeat. Perform three sets of 15 repetitions.

AVOID
- Arching your back.
- Flaring your elbows outward.
- Swinging your weights—especially important as the weights are close to your head.

ANNOTATION KEY

Black text indicates target muscles

Gray text indicates other working muscles

* indicates deep muscles

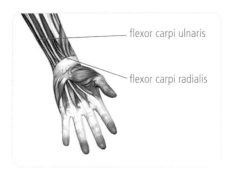

flexor carpi ulnaris

flexor carpi radialis

triceps brachii

deltoideus anterior

pectoralis major

latissimus dorsi

teres major

deltoideus posterior

SWISS BALL FLY

❶ Lie face-up on a Swiss ball, with your upper back, neck, and head supported. Your body should be extended with your torso long, knees bent at a right angle, and feet planted on the floor a little wider than shoulder-distance apart. Grasp a hand weight or dumbbell in each hand and extend your arms straight up.

DO IT RIGHT
- When lifting the weights overhead, keep your arms directly above your shoulders.
- Keep your torso stable and feet planted throughout the exercise.
- Engage your abs.
- Keep your buttocks and pelvis lifted so that your upper legs, torso, and neck form a straight line.
- Move your arms smoothly and with control.

AVOID
- Arching your back.
- Swinging your arms.

TARGETS
- Chest

LEVEL
- Beginner

BENEFITS
- Strengthens and tones pectoral muscles

NOT ADVISABLE IF YOU HAVE . . .
- Shoulder issues

❷ Keeping the rest of your body stable, bring your arms to your sides.

❸ Return your arms to starting position. Repeat, completing three sets of 15.

MODIFICATION
Similar level of difficulty: Instead of holding hand weights, loop a fitness band under your ball and grasp both handles. Keep your arms extended as you hold the strap taut throughout the exercise.

ANNOTATION KEY

Black text indicates target muscles

Gray text indicates other working muscles

* indicates deep muscles

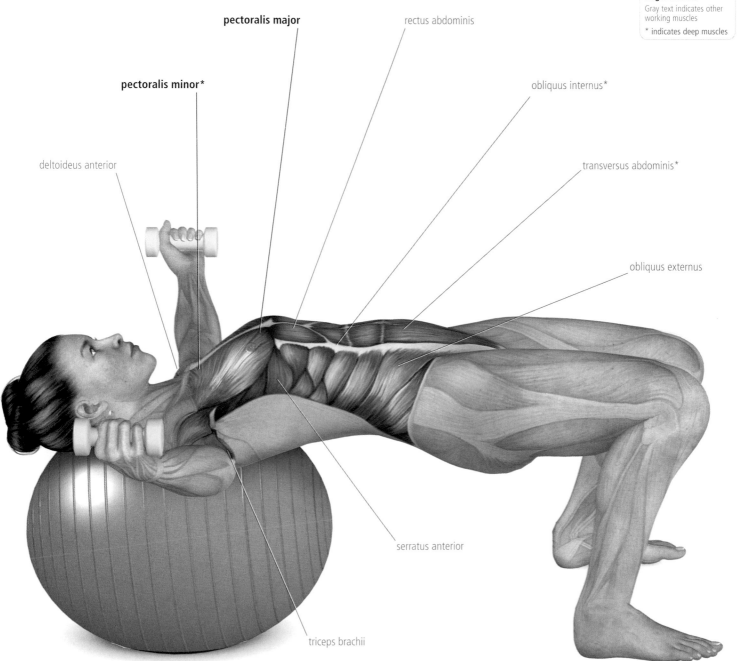

pectoralis major

rectus abdominis

pectoralis minor*

obliquus internus*

deltoideus anterior

transversus abdominis*

obliquus externus

serratus anterior

triceps brachii

PUSH-UP

❶ From a standing position, walk your hands out until they are directly beneath your shoulders in a high plank position.

❷ Inhale, and set your body by drawing your abdominals to your spine. Squeeze your buttocks and legs together and stretch out of your heels, bringing your body into a straight line.

❸ Exhale and inhale as you bend your elbows and lower your body toward the floor.

TARGETS
• Chest
• Upper arms

LEVEL
• Beginner

BENEFITS
• Strengthens the core stabilizers, shoulders, back, buttocks, and pectoral muscles

NOT ADVISABLE IF YOU HAVE . . .
• Shoulder issues
• Wrist pain
• Lower-back pain

❸ Push upward to return to plank position. Keep your elbows close to your body. Repeat eight times.

❹ Inhale as you lift your hips into the air, and walk your hands back toward your feet. Exhale slowly, rolling up one vertebra at a time into your starting position. Repeat the entire exercise three times.

DO IT RIGHT
- Relax your neck, keeping it long as you perform the upward movement.
- Squeeze your glutes as you scoop in your abdominals for stability.

AVOID
- Allowing your shoulders to lift toward your ears.

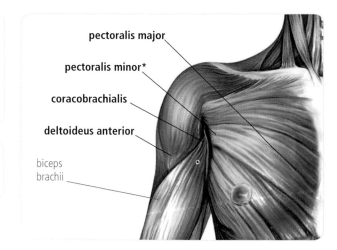

pectoralis major

pectoralis minor*

coracobrachialis

deltoideus anterior

biceps brachii

BEST FOR

- triceps brachii
- pectoralis major
- pectoralis minor
- coracobrachialis
- deltoideus anterior
- rectus abdominis
- transversus abdominis
- obliquus externus
- obliquus internus
- trapezius

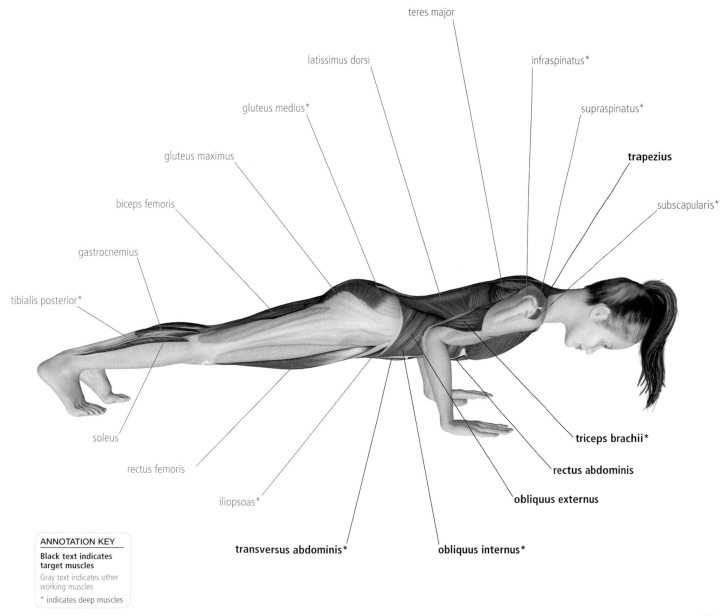

teres major

latissimus dorsi

gluteus medius*

gluteus maximus

biceps femoris

gastrocnemius

tibialis posterior*

soleus

rectus femoris

iliopsoas*

transversus abdominis*

obliquus internus*

infraspinatus*

supraspinatus*

trapezius

subscapularis*

triceps brachii*

rectus abdominis

obliquus externus

ANNOTATION KEY

Black text indicates target muscles

Gray text indicates other working muscles

* indicates deep muscles

PRONE TRUNK RAISE

1 Lie prone on the floor. Bend your elbows, placing your hands flat on the floor on either side of your chest. Keep your elbows pulled in toward your body. Separate your legs one hip-width apart, and extend through your toes. The tops of your feet should be touching the floor.

2 Inhale, and press against the floor with your hands and the tops of your feet, lifting your torso and hips off the floor. Contract your thighs, and tuck your tailbone toward your pubis.

3 Lift through the top of your chest, fully extending your arms and creating an arch in your back from your upper torso. Push your shoulders down and back, and elongate your neck as you gaze slightly upward.

4 Hold for 15 to 30 seconds, and exhale as you lower yourself to the floor.

TARGETS
• Upper back
• Lower back
• Upper arms
• Gluteal muscles

LEVEL
• Beginner

BENEFITS
• Strengthens spine, arms, and wrists
• Stretches chest and abdominals
• Improves posture

NOT ADVISABLE IF YOU HAVE . . .
• Back injury
• Wrist injury or carpal tunnel syndrome

DO IT RIGHT
• Elongate your legs and arms to create full extension.
• Make sure that your wrists are positioned directly below your shoulders so that you don't exert too much pressure on your lower back.

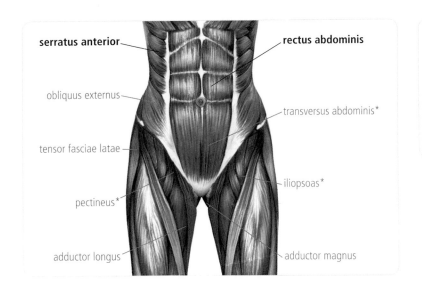

serratus anterior

rectus abdominis

obliquus externus

transversus abdominis*

tensor fasciae latae

iliopsoas*

pectineus*

adductor longus

adductor magnus

AVOID
- Lifting your shoulders up toward your ears.
- Hyperextending your elbows.
- Jutting your rib cage out of your chest.
- Dropping your thighs to the floor.

BEST FOR
- rhomboideus
- teres major
- teres minor
- trapezius
- latissimus dorsi
- erector spinae
- quadratus lumborum
- gluteus maximus
- pectoralis major
- serratus anterior
- rectus abdominis
- triceps brachii

ANNOTATION KEY
Black text indicates target muscles
Gray text indicates other working muscles
* indicates deep muscles

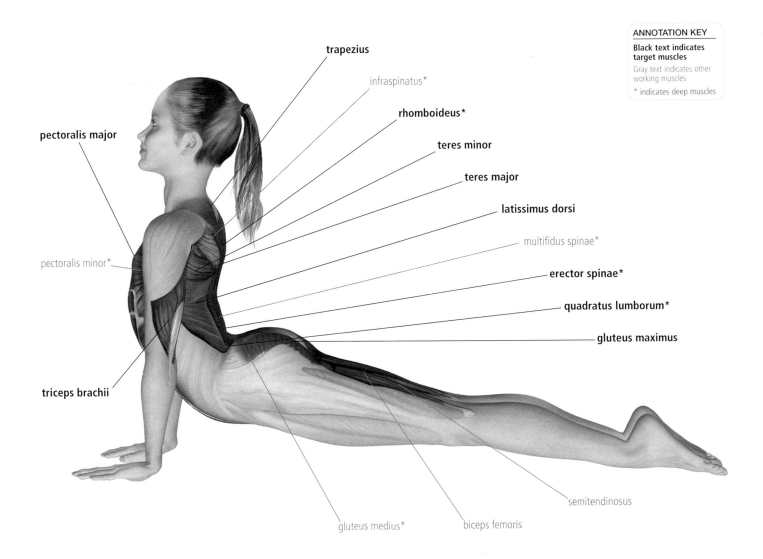

trapezius

infraspinatus*

rhomboideus*

teres minor

teres major

pectoralis major

latissimus dorsi

multifidus spinae*

pectoralis minor*

erector spinae*

quadratus lumborum*

gluteus maximus

triceps brachii

semitendinosus

gluteus medius*

biceps femoris

DUMBBELL UPRIGHT ROW

1 Stand with your feet parallel and shoulder-width apart, holding a pair of dumbbells in front of your thighs.

2 Bend your elbows to the side as you raise your weights, aiming for shoulder height.

3 Lower the dumbbells to starting position. Repeat, completing three sets of 15.

BACK VIEW

TARGETS
• Shoulders
• Upper back

LEVEL
• Beginner

BENEFITS
• Strengthens muscles in upper back and shoulders

NOT ADVISABLE IF YOU HAVE . . .
• Shoulder issues
• Tennis elbow

AVOID
- Swinging your weights; instead, move slowly and with control.
- Arching your back or slump forward.

DO IT RIGHT
- Keep your torso stable, your back straight, and your abs engaged.
- Lead with your elbows.

ANNOTATION KEY

Black text indicates target muscles

Gray text indicates other working muscles

* indicates deep muscles

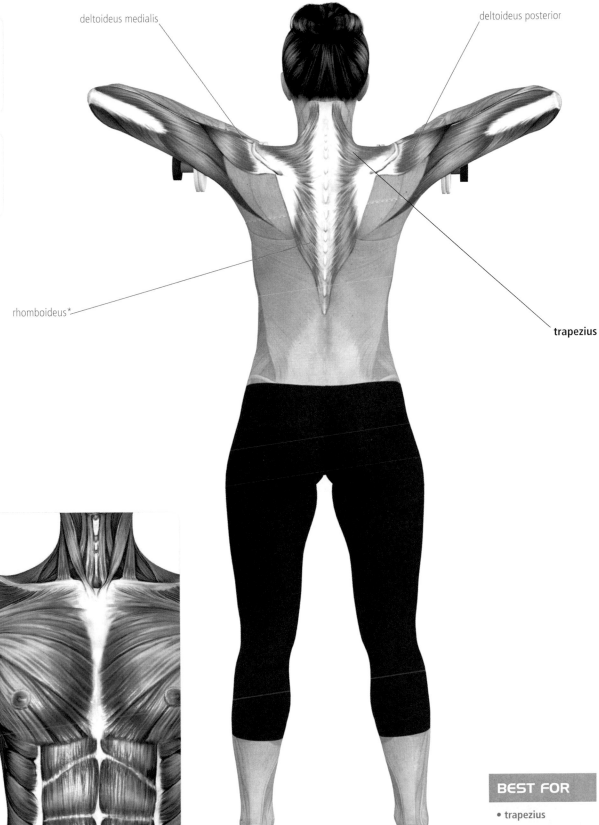

deltoideus medialis

deltoideus posterior

rhomboideus*

trapezius

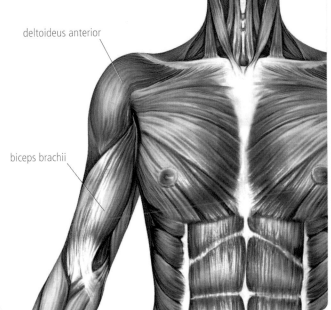

deltoideus anterior

biceps brachii

BEST FOR
- trapezius

ALTERNATING DUMBBELL CURL

1 Stand upright, with your feet planted about shoulder-width apart and your knees very slightly bent. Hold a hand weight or dumbbell in each hand, with your arms down along your sides, palms facing forward.

2 In a smooth, controlled movement, bend one arm as you raise the weight toward your shoulder.

3 As you begin to lower your arm, begin to raise the other one, and repeat on the other side. Continue to alternate, completing three sets of 15 per arm.

TARGETS
• Biceps

LEVEL
• Beginner

BENEFITS
• Strengthens and tones upper arms

NOT ADVISABLE IF YOU HAVE . . .
• Lower-back issues

DO IT RIGHT
• Keep your knees soft throughout the exercise.
• Gaze forward.
• Keep one arm still while the other is moving.
• Keep your torso still.

ANNOTATION KEY

Black text indicates target muscles

Gray text indicates other working muscles

* indicates deep muscles

BEST FOR

• biceps brachii

levator scapulae*

deltoideus anterior

trapezius

biceps brachii

brachialis

brachioradialis

flexor carpi ulnaris

flexor carpi radialis

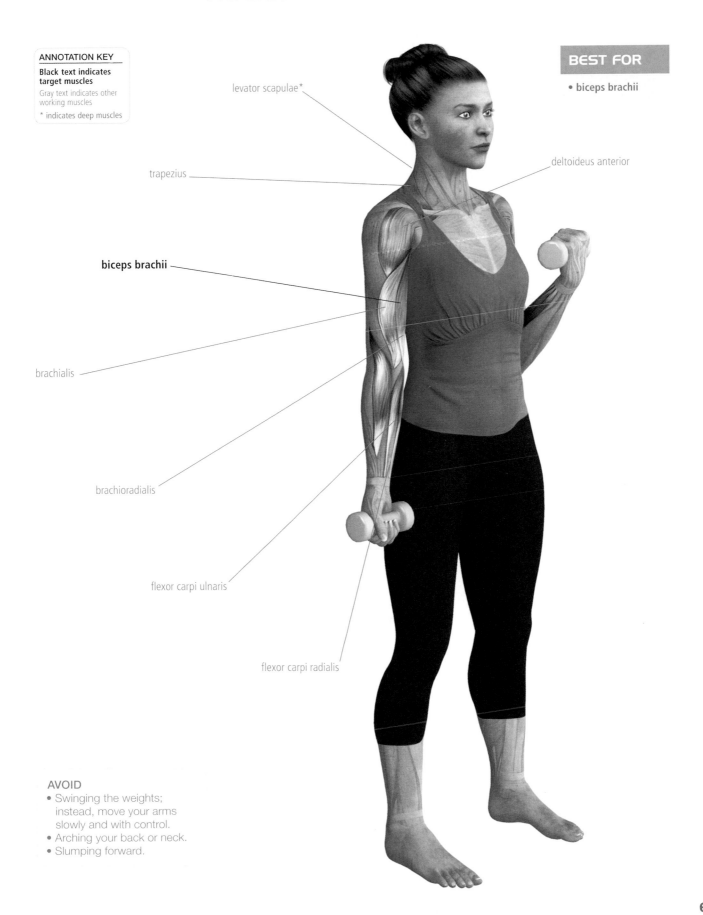

AVOID

• Swinging the weights; instead, move your arms slowly and with control.

• Arching your back or neck.

• Slumping forward.

CORE-TRAINING EXERCISES

A sleek, toned midsection and a well-defined waist are central

to the notion of feminine beauty, and most of us begin our

exercise programs with the goal to tone these areas. But a

strong and stable core doesn't just make you look fitter and

sexier, it is (quite literally) central to how you function. All

of your body's movements, in every conceivable direction,

originate in the core, and when you strengthen it, you guard

against injury, improve functionality and posture, and build

fitness from the inside out. The core is the powerhouse of

the body, working so that you can lift, bend, and carry all of

your daily burdens, from kids to grocery bags. The following

exercises train the major core muscles, including the abs, the

obliques, and the muscles that support the spine.

CRUNCH

1 Lie on your back with your knees bent, and clasp your hands behind your head.

2 Keeping your elbows wide, engage your abdominals, and lift your upper torso to achieve a crunching movement.

3 Slowly return to the starting position. Repeat 15 times for two sets.

TARGETS
• Abdominals

LEVEL
• Beginner

BENEFITS
• Strengthens the torso
• Improves pelvic and core stability

NOT ADVISABLE IF YOU HAVE . . .
• Back pain
• Neck pain

MODIFICATION

Harder: Lie on your back with your legs outstretched, and your arms over your head. Without lifting your legs, lift your arms and torso in a controlled movement. Continue to curl forward and grasp your feet.

DO IT RIGHT
• Use your shoulders and abdominals to initiate the movement.
• Keep your pelvis in neutral position during the crunching motion.
• Slightly tuck your chin, directing your gaze toward the inner thighs.

AVOID
• Pulling from the neck.
• Tilting your hips toward the floor.

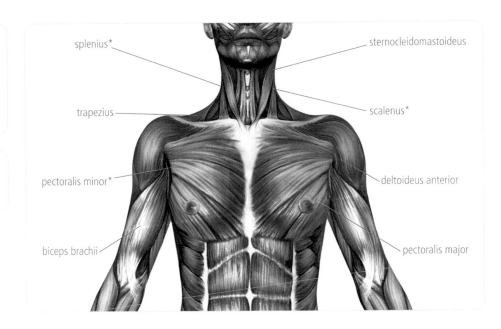

splenius*
sternocleidomastoideus
trapezius
scalenus*
pectoralis minor*
deltoideus anterior
biceps brachii
pectoralis major

ANNOTATION KEY
Black text indicates target muscles
Gray text indicates other working muscles
* indicates deep muscles

BEST FOR
• rectus abdominis
• obliquus internus
• obliquus externus
• transversus abdominis

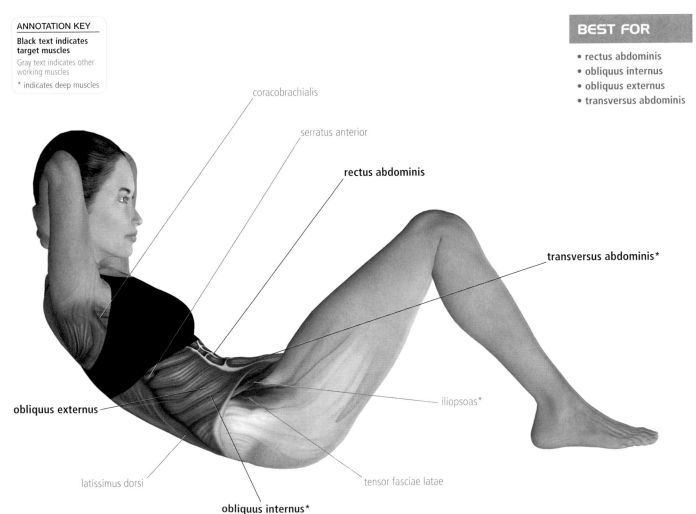

coracobrachialis
serratus anterior
rectus abdominis
transversus abdominis*
obliquus externus
iliopsoas*
latissimus dorsi
tensor fasciae latae
obliquus internus*

HALF CURL

❶ Lie on your back with your knees bent and arms straight by your sides. Squeeze your legs together and keep your feet flat on the floor.

AVOID
- Curling your neck too far forward.
- Allowing your feet to raise off the floor.
- Raising up too far.

TARGETS
- Upper abdominals

LEVEL
- Beginner

BENEFITS
- Strengthens core muscles
- Increases abdominal endurance

NOT ADVISABLE IF YOU HAVE . . .
- Neck issues

❷ Using your upper abdominals, curl your upper back and shoulders upward. Keep your arms parallel to the floor and your lower back flat.

❸ Hold for 10 seconds. Return to the starting position, and repeat 10 times.

DO IT RIGHT
- Keep your arms parallel to the floor.

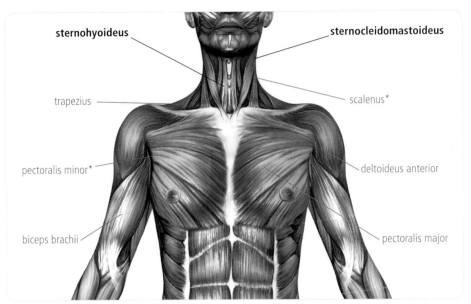

sternohyoideus

sternocleidomastoideus

trapezius

scalenus*

pectoralis minor*

deltoideus anterior

biceps brachii

pectoralis major

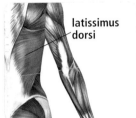

latissimus dorsi

ANNOTATION KEY

Black text indicates target muscles

Gray text indicates other working muscles

* indicates deep muscles

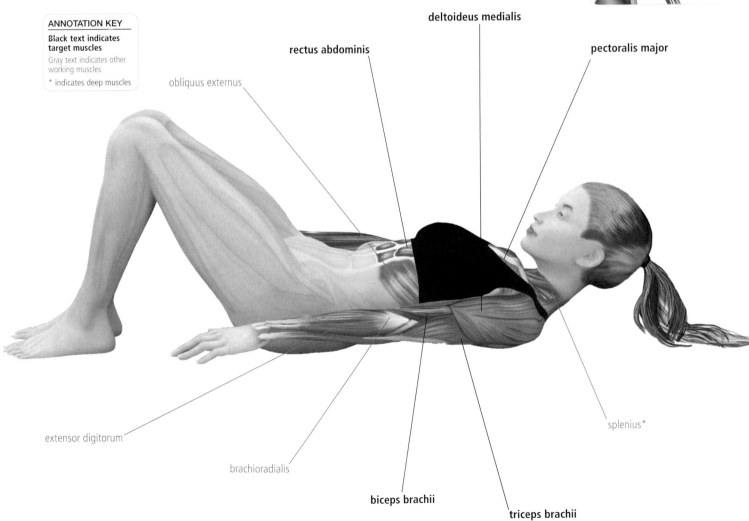

deltoideus medialis

rectus abdominis

pectoralis major

obliquus externus

splenius*

extensor digitorum

brachioradialis

biceps brachii

triceps brachii

SEATED RUSSIAN TWIST

❶ Sit upright with your legs bent, feet flat on the floor. Extend your arms straight ahead, and lean back slightly to activate your core.

DO IT RIGHT
• Twist smoothly and with control.
• Keep your back flat as you twist.
• Keep your feet on the floor.
• Keep your arms straight.

AVOID
• Rushing through the twist.
• Shifting your feet or knees to the side as you twist.

TARGETS
• Back
• Obliques
• Upper abdominals

LEVEL
• Intermediate

BENEFITS
• Stabilizes and strengthens core

NOT ADVISABLE IF YOU HAVE . . .
• Lower-back pain

❷ In a smooth motion, rotate your upper body to the side, and then return to center. Repeat rotation on the other side.

❸ Return to center, and repeat the full twist, performing three sets of 20.

BEST FOR

• rectus abdominis
• obliquus externus
• obliquus internus
• erector spinae
• transversus abdominis

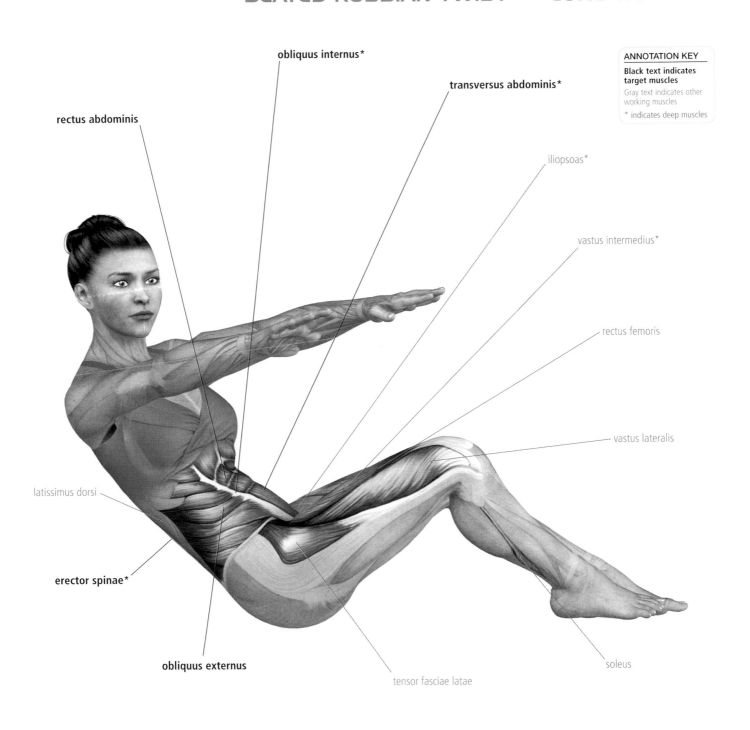

obliquus internus*

transversus abdominis*

rectus abdominis

ANNOTATION KEY

Black text indicates
target muscles

Gray text indicates other
working muscles

* indicates deep muscles

iliopsoas*

vastus intermedius*

rectus femoris

vastus lateralis

latissimus dorsi

erector spinae*

obliquus externus

soleus

tensor fasciae latae

MODIFICATION

Harder: Perform twists
holding a medicine ball.

SPINE TWIST

1 Sit on the floor, with your back straight. Extend your legs in front of you, slightly more than hip-width apart.

2 Lift yourself as tall as you can from the base of your spine. Ground your hips into the floor.

AVOID
- Allowing your hips to rise off the floor.

DO IT RIGHT
- Rotate your torso along the central axis of your body.
- Keep your arms parallel to the floor.
- Keep your back straight; if your hamstrings are too tight to allow you to sit up straight, place a towel under your buttocks, and bend your knees slightly.

TARGETS
- Spine

LEVEL
- Beginner

BENEFITS
- Strengthens and lengthens the torso

NOT ADVISABLE IF YOU HAVE . . .
- Back pain

3 Lift up and out of your hips as you pull in your lower abdominals. Twist from your waist to the left, keeping your hips squared and grounded.

4 Slowly return to the center.

5 Lift up and out of your hips again, twisting in the other direction.

6 Return to the center. Repeat three times in each direction.

BEST FOR

- transversus abdominis
- obliquus externus
- biceps femoris
- gluteus maximus
- tensor fasciae latae
- latissimus dorsi
- teres major
- quadratus lumborum
- deltoideus posterior
- rectus femoris

ANNOTATION KEY

Black text indicates target muscles

Gray text indicates other working muscles

* indicates deep muscles

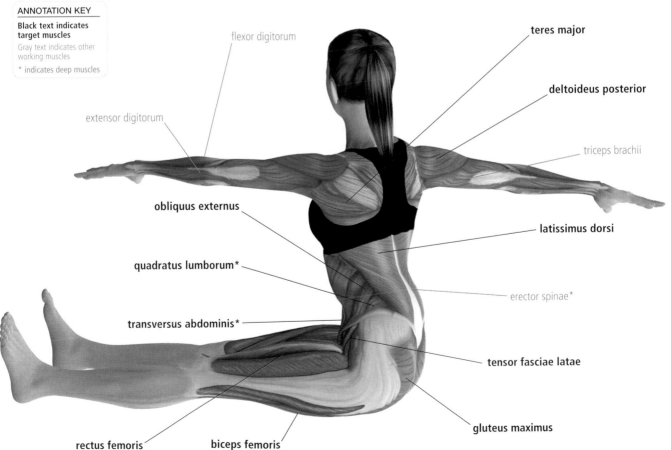

flexor digitorum

teres major

deltoideus posterior

extensor digitorum

triceps brachii

obliquus externus

latissimus dorsi

quadratus lumborum*

erector spinae*

transversus abdominis*

tensor fasciae latae

gluteus maximus

rectus femoris

biceps femoris

79

OBLIQUE ROLL-DOWN

1 Sit with your knees bent and your arms extended to the sides, parallel to the floor.

2 Contract your abdominals, drawing your navel to your spine and lengthening the spine upward.

TARGETS
• Obliques

LEVEL
• Advanced

BENEFITS
• Tightens the obliques and abdominals

NOT ADVISABLE IF YOU HAVE . . .
• Herniated disk

3 Roll backward while simultaneously rotating your torso to one side.

4 Maintaining spinal flexion, rotate your torso back to the center.

AVOID
• Tensing your neck and shoulder muscles.

5 Rotate to the other side, deepening the abdominal contraction.

6 Return back to the center, and repeat sequence four to six times on each side.

DO IT RIGHT
- Lengthen your arms as you roll down to create opposition throughout the torso.
- Relax and lengthen your neck to prevent straining.
- Articulate your spine while rolling up and down.

ANNOTATION KEY

Black text indicates target muscles

Gray text indicates other working muscles

* indicates deep muscles

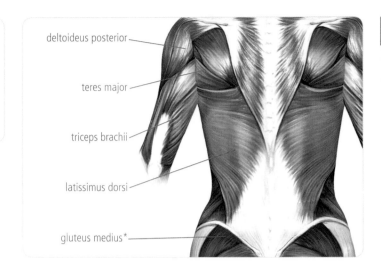

deltoideus posterior
teres major
triceps brachii
latissimus dorsi
gluteus medius*

BEST FOR
- obliquus externus
- obliquus internus
- rectus abdominis
- transversus abdominis

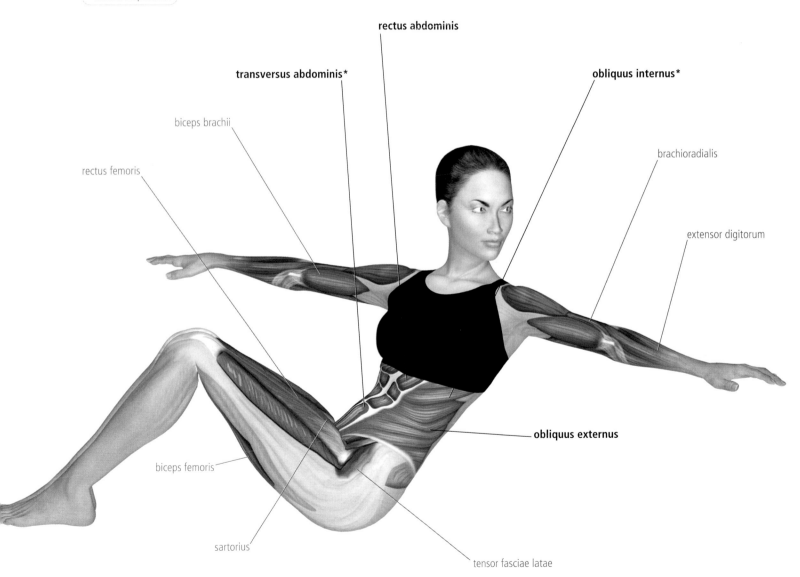

rectus abdominis

transversus abdominis*

biceps brachii

rectus femoris

obliquus internus*

brachioradialis

extensor digitorum

obliquus externus

biceps femoris

sartorius

tensor fasciae latae

BICYCLE CRUNCH

① Lie on your back with your fingers at your ears, your elbows flared outward, and your legs bent at a 90-degree angle.

DO IT RIGHT
- Use your core to drive the movement.
- Keep your elbows flared.
- Keep both hips stable on the floor.
- Keep your neck elongated.

② Roll up with your torso, reaching one elbow diagonally toward the opposite knee. At the same time, extend the other leg forward.

TARGETS
- Obliques
- Upper abdominals

LEVEL
- Intermediate

BENEFITS
- Stabilizes core
- Strengthens and tones obliques and upper abdominals
- Hip flexors

NOT ADVISABLE IF YOU HAVE . . .
- Neck issues
- Lower-back pain

③ Release, and repeat on the other side. Continue to alternate, completing 15 crunches in each direction.

BEST FOR
- rectus abdominis
- obliquus internus
- obliquus externus

MODIFICATION
Similar level of difficulty:
Keep both feet flexed throughout the exercise.

AVOID
• Arching your back or raise your lower back off the floor.
• Pulling your head upward with your hands.

ANNOTATION KEY

Black text indicates target muscles

Gray text indicates other working muscles

* indicates deep muscles

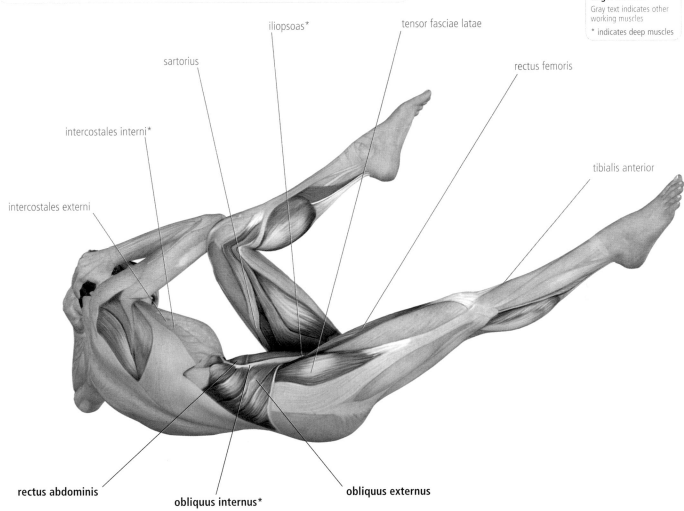

sartorius

iliopsoas*

tensor fasciae latae

rectus femoris

intercostales interni*

tibialis anterior

intercostales externi

rectus abdominis

obliquus internus*

obliquus externus

MODIFICATION
Easier: Begin with one foot on the floor, and place the outside of your other foot on top of your thigh near your knee. As you crunch, bring your opposite elbow toward that top knee. Complete five reps on one side, then switch sides and repeat.

THE BOAT

1. Sit with your legs extended straight in front of you.

2. Lean back slightly, bending your knees, and support yourself with your hands behind your hips. Your fingers should be pointing forward, and your back should be straight.

DO IT RIGHT
- Keep your neck elongated and relaxed, minimizing the tension in your upper spine.
- If you are unable to straighten your legs, balance with your knees bent.

3. Exhale, and lift your feet off the floor as you lean back from your shoulders. Find your balance point between your sit bones and your tailbone.

4. Slowly straighten your legs in front of you so that they form a 45-degree angle with your torso. Point your toes. Lift your arms to your sides, parallel to the floor.

5. Pull your abdominals in toward your spine as they work to keep your balance. Stretch your arms forward through your fingertips, and elongate the back of your neck.

6. Hold for 10 to 20 seconds.

TARGETS
- Abdominals
- Hip flexors

LEVEL
- Advanced

BENEFITS
- Strengthens abdominals, hip flexors, spine, and thighs
- Stretches hamstrings

NOT ADVISABLE IF YOU HAVE . . .
- Neck injury
- Headache
- Lower-back pain

BEST FOR

- rectus abdominis
- obliquus internus
- obliquus externus
- iliopsoas
- transversus abdominis
- vastus intermedius
- rectus femoris
- erector spinae

AVOID
- Rounding your spine, which places pressure on your lower back.

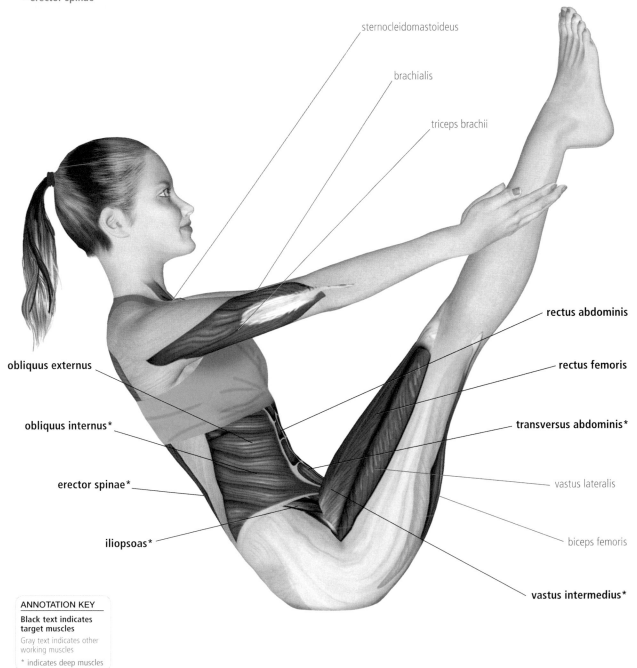

sternocleidomastoideus

brachialis

triceps brachii

rectus abdominis

rectus femoris

transversus abdominis*

vastus lateralis

biceps femoris

vastus intermedius*

obliquus externus

obliquus internus*

erector spinae*

iliopsoas*

ANNOTATION KEY

Black text indicates target muscles

Gray text indicates other working muscles

* indicates deep muscles

V-UP

❶ Lie on your back with your legs raised a few inches from the floor.

❷ Inhale, reaching your arms toward the ceiling as you lift your head and shoulders off the floor.

AVOID
- Using momentum to carry you through the exercise; instead, use your abdominal muscles to lift your legs and torso.

TARGETS
- Abdominals

LEVEL
- Advanced

BENEFITS
- Strengthens the abdominals
- Increases spinal flexibility

NOT ADVISABLE IF YOU HAVE . . .
- Advanced osteoporosis
- Herniated disk

❸ Exhale, and keeping them straight, begin lifting your legs to a 45-degree angle from the floor.

DO IT RIGHT
- Articulate through the spine on the way up and on the way down.
- Keep your neck elongated and relaxed, minimizing the tension in your upper spine.

④ While rolling through the spine, lift your rib cage off the floor to just before the sit bones.

⑤ Inhale, and reach your arms toward your toes while maintaining a C curve in your back. Exhale, and roll down the spine by articulating one vertebra at a time. Return to the starting position.

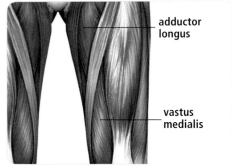

adductor longus

vastus medialis

BEST FOR

- rectus abdominis
- tensor fasciae latae
- rectus femoris
- vastus lateralis
- vastus medialis
- vastus intermedius
- adductor longus
- pectineus
- brachialis

ANNOTATION KEY

Black text indicates target muscles

Gray text indicates other working muscles

* indicates deep muscles

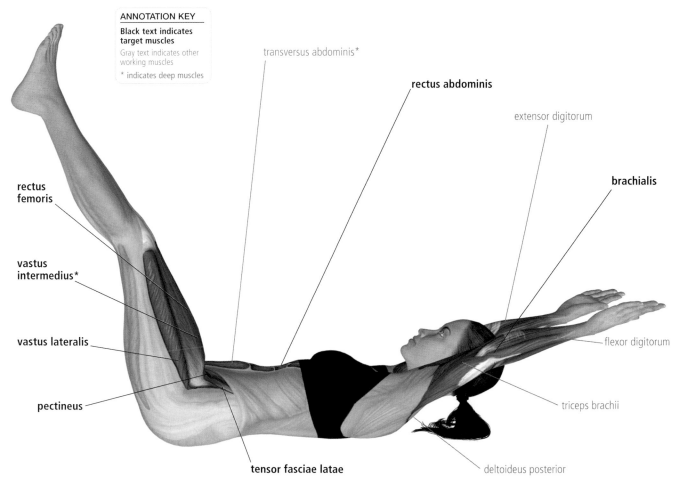

rectus femoris

vastus intermedius*

vastus lateralis

pectineus

tensor fasciae latae

transversus abdominis*

rectus abdominis

extensor digitorum

brachialis

flexor digitorum

triceps brachii

deltoideus posterior

BACKWARD BALL STRETCH

DO IT RIGHT
- Maintain good balance throughout the stretch.
- Move slowly and in a controlled manner.
- Keep your head on the ball until you have dropped your knees all the way down as you release from the stretch.

AVOID
- Allowing the ball to shift to the side.
- Holding the extended position for too long, or until you feel dizzy.

TARGETS
- Thoracic and upper-lumbar spine
- Abdominals

LEVEL
- Advanced

BENEFITS
- Stretches thoracic spine
- Increases spinal extension
- Stretches abdominals and large back muscles

NOT ADVISABLE IF YOU HAVE . . .
- Lower-back pain
- Balancing difficulty

❶ Sit on a Swiss ball in a well-balanced, neutral position, with your hips directly over the center of the ball.

❷ Raise your arms while maintaining good balance, and begin to extend them behind you.

❸ As you continue to extend your hands backward, walk your feet forward, allowing the ball to roll up your spine.

❹ As your hands touch the floor, extend your legs as far forward as you comfortably can. Hold this position for 10 seconds.

❺ To deepen the stretch, extend your arms, and walk your legs and hands closer to the ball. Hold this position for 10 seconds.

❻ To release the stretch, bend your knees, drop your hips to the floor, lift your head off the ball, and then walk back to the starting position.

MODIFICATION

Easier: Follow steps 1 through 3, but rather than extend your hands to the floor, clasp them behind your head. Hold this position for 10 seconds, and release.

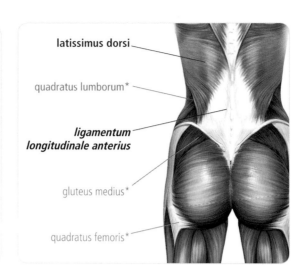

latissimus dorsi

quadratus lumborum*

ligamentum longitudinale anterius

gluteus medius*

quadratus femoris*

BEST FOR

- deltoideus medialis
- iliopsoas
- latissimus dorsi
- serratus anterior
- pectoralis major
- pectoralis minor
- ligamentum longitudinale anterius

ANNOTATION KEY

Black text indicates target muscles

Gray text indicates other working muscles

Black italics indicates ligaments

* indicates deep muscles

rectus abdominis

obliquus externus

transversus abdominis*

vastus lateralis

rectus femoris

biceps femoris

iliopsoas*

serratus anterior

pectoralis major

pectoralis minor*

deltoideus medialis

trapezius

biceps brachii

flexor carpi radialis

PLANK

❶ Lie on your stomach, with your legs extended behind you. Bend your arms so that your forearms and palms rest flat on the floor.

TARGETS
- Abdominals
- Back
- Obliques

LEVEL
- Beginner

BENEFITS
- Strengthens and stabilizes core

NOT ADVISABLE IF YOU HAVE . . .
- Shoulder injury
- Severe back pain

❷ Bend your knees, supporting your weight between your knees and your forearms, and then push through with your forearms to bring your shoulders up toward the ceiling as you straighten your legs.

❸ With control, lower your shoulders until you feel them coming together at your back. Hold for 30 seconds, building up to 2 minutes if desired.

BEST FOR

- erector spinae
- transversus abdominis
- rectus abdominis
- obliquus externus
- obliquus internus

AVOID
- Allowing your shoulders to collapse into your shoulder joints.
- Arching your neck.
- Allowing your back to sag.

DO IT RIGHT
- Keep your abs tight.
- Keep your body in a straight line.
- Lengthen through your neck.
- Start by holding for just 15 seconds, if desired.

ANNOTATION KEY
Black text indicates target muscles
Gray text indicates other working muscles
* indicates deep muscles

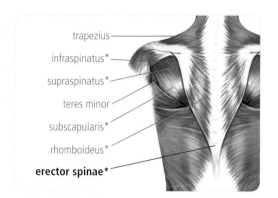

- trapezius
- infraspinatus*
- supraspinatus*
- teres minor
- subscapularis*
- rhomboideus*
- **erector spinae***

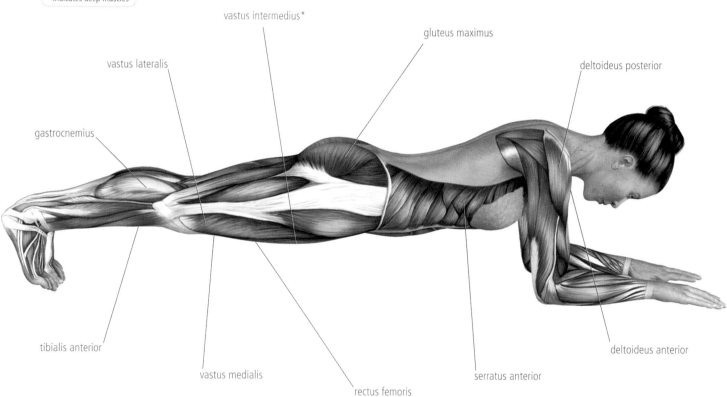

- vastus intermedius*
- gluteus maximus
- vastus lateralis
- deltoideus posterior
- gastrocnemius
- tibialis anterior
- vastus medialis
- rectus femoris
- serratus anterior
- deltoideus anterior

MODIFICATION
Harder: While in the plank position, lift and lower your legs one at a time. Keep the rest of your body still, and your abs engaged throughout.

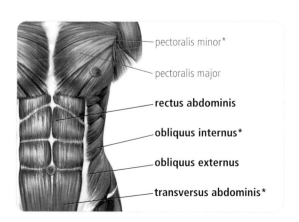

- pectoralis minor*
- pectoralis major
- **rectus abdominis**
- **obliquus internus***
- **obliquus externus**
- **transversus abdominis***

SWISS BALL TRANSVERSE ABS

1 Position yourself on your toes with your arms bent and forearms resting on top of a Swiss ball.

2 Form a long, straight line from your ankles to your shoulders.

3 Hold this position for as long as you can.

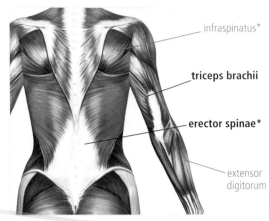

infraspinatus*

triceps brachii

erector spinae*

extensor digitorum

TARGETS
• Lower abdominals
• Upper back

LEVEL
• Advanced

BENEFITS
• Stabilizes core
• Strengthens abdominals
• Strengthens lower back

NOT ADVISABLE IF YOU HAVE . . .
• Neck pain
• Lower-back pain

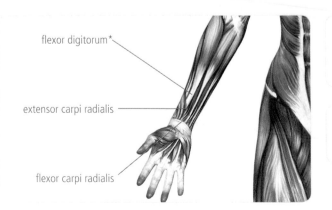

flexor digitorum*

extensor carpi radialis

flexor carpi radialis

DO IT RIGHT
• Breathe easily and normally.
• Activate your abs so that you maintain neutral alignment.

AVOID
• Allowing your lower back to drop out of alignment.

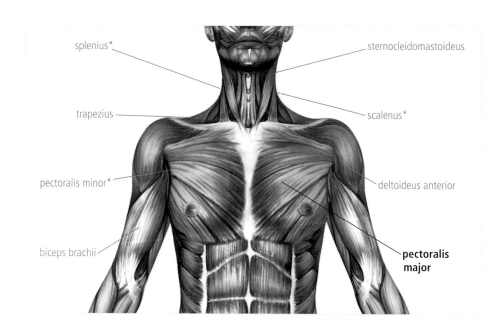

splenius*

sternocleidomastoideus

trapezius

scalenus*

pectoralis minor*

deltoideus anterior

biceps brachii

pectoralis major

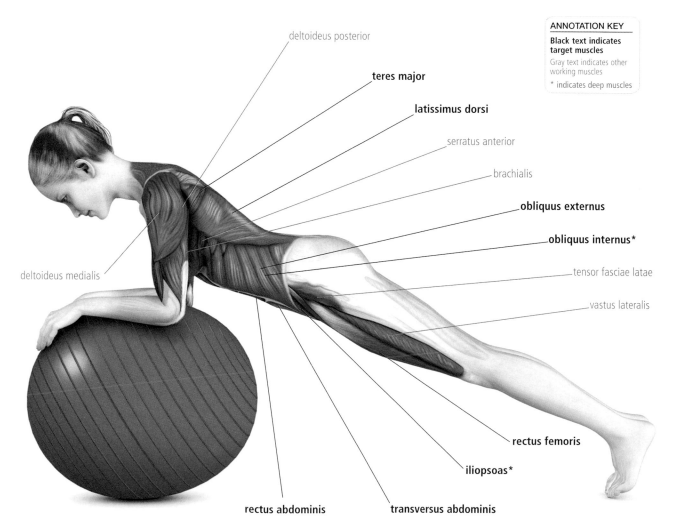

deltoideus posterior

teres major

latissimus dorsi

serratus anterior

brachialis

obliquus externus

obliquus internus*

tensor fasciae latae

vastus lateralis

deltoideus medialis

rectus femoris

iliopsoas*

rectus abdominis

transversus abdominis

ANNOTATION KEY

Black text indicates target muscles

Gray text indicates other working muscles

* indicates deep muscles

SWISS BALL ROLLOUT

BEST FOR

• rectus abdominis
• erector spinae

1 Kneel in front of a Swiss ball, with your hands resting on the ball.

DO IT RIGHT
• Keep your upper body elongated.
• Keep your lower legs and feet anchored to the floor throughout the exercise.
• Maintain a flat back.
• Keep your abs pulled in.
• Move smoothly and with control.

TARGETS
• Back
• Upper abdominals

LEVEL
• Intermediate

BENEFITS
• Stabilizes core

**NOT ADVISABLE
IF YOU HAVE . . .**
• Lower-back issues
• Knee Issues

2 Use your hands to roll the ball slightly in front of you as you begin to lean forward.

AVOID
• Allowing your hips to sag.

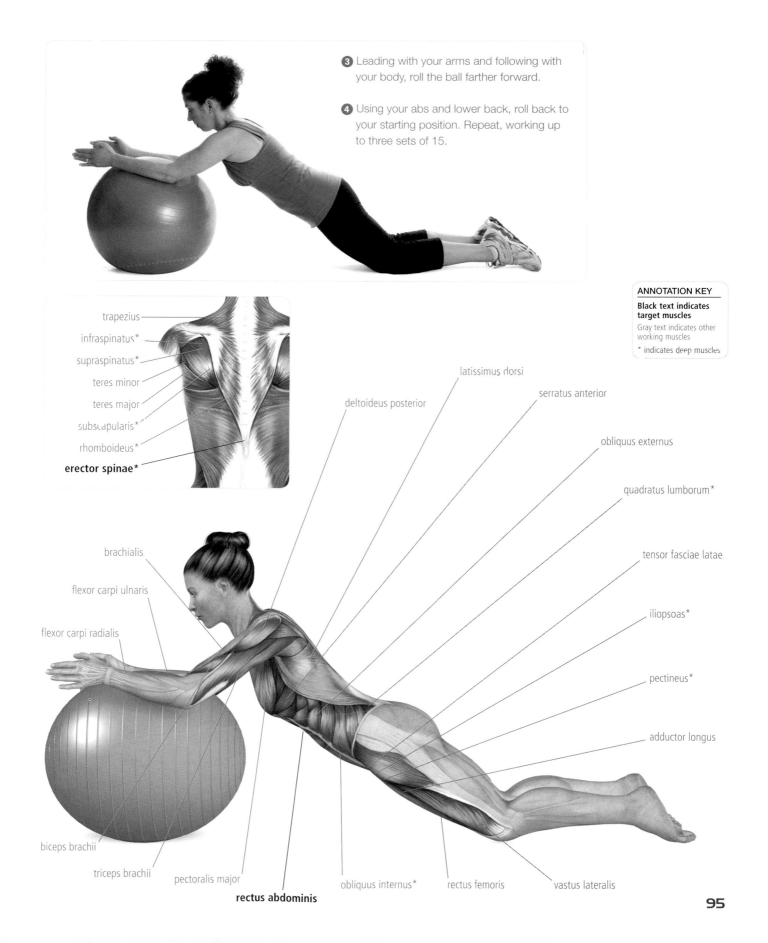

3 Leading with your arms and following with your body, roll the ball farther forward.

4 Using your abs and lower back, roll back to your starting position. Repeat, working up to three sets of 15.

ANNOTATION KEY

Black text indicates target muscles

Gray text indicates other working muscles

* indicates deep muscles

trapezius

infraspinatus*

supraspinatus*

teres minor

teres major

subscapularis*

rhomboideus*

erector spinae*

latissimus dorsi

deltoideus posterior

serratus anterior

obliquus externus

quadratus lumborum*

tensor fasciae latae

iliopsoas*

pectineus*

adductor longus

brachialis

flexor carpi ulnaris

flexor carpi radialis

biceps brachii

triceps brachii

pectoralis major

obliquus internus*

rectus femoris

vastus lateralis

rectus abdominis

FOAM ROLLER CALF PRESS

DO IT RIGHT
- Form a long, straight line with your lifted leg.
- Keep your hips elevated throughout the exercise.

AVOID
- Allowing your shoulders to lift toward your ears.
- Bending your knees.
- Bending your elbows.

1 Sit with your legs outstretched in front of you, with a foam roller placed under your knees. Place your hands on the floor to support your torso, your fingers pointing toward your buttocks.

TARGETS
- Abdominals
- Upper arms
- Shoulder stabilizers
- Hamstrings

LEVEL
- Advanced

BENEFITS
- Improves core, pelvic, and shoulder stability

NOT ADVISABLE IF YOU HAVE . . .
- Wrist pain
- Shoulder pain
- Knee issues

2 Press into the floor to lift your hips, keeping your legs firm.

BEST FOR
- rectus abdominis
- transversus abdominis
- triceps brachii
- serratus anterior
- deltoideus anterior
- biceps femoris
- semitendinosus
- semimembranosus

3 Lift one leg off the roller and hold it steady, making sure not to drop your hips.

4 Keep your leg lifted, and press your opposite leg into the roller, drawing your hips back toward your hands.

5 Return to the starting position, rolling your calf muscle along the roller and keeping your lifted leg straight in the air. Repeat 15 times on each leg.

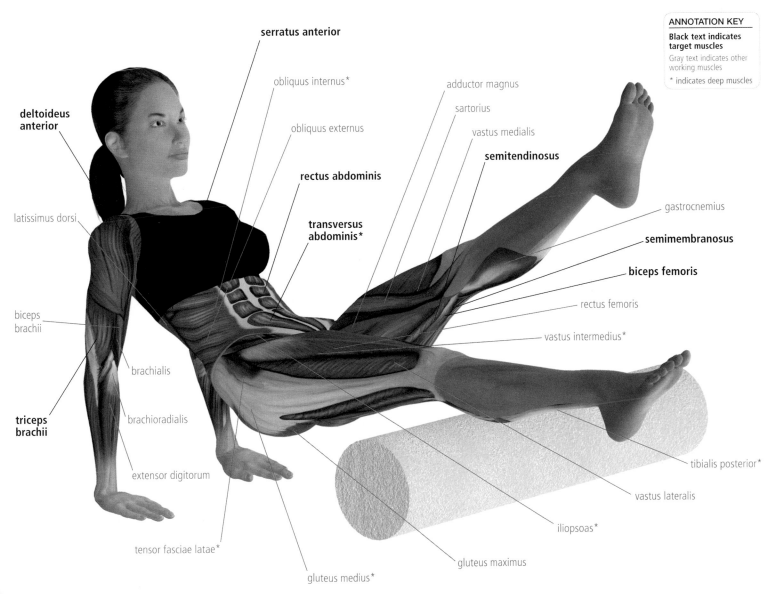

ANNOTATION KEY

Black text indicates target muscles

Gray text indicates other working muscles

* indicates deep muscles

serratus anterior

obliquus internus*

obliquus externus

rectus abdominis

transversus abdominis*

deltoideus anterior

latissimus dorsi

biceps brachii

brachialis

triceps brachii

brachioradialis

extensor digitorum

tensor fasciae latae*

gluteus medius*

gluteus maximus

iliopsoas*

vastus lateralis

tibialis posterior*

biceps femoris

rectus femoris

vastus intermedius*

semimembranosus

gastrocnemius

semitendinosus

adductor magnus

sartorius

vastus medialis

FOAM ROLLER DIAGONAL CRUNCH

❶ Lie lengthwise on a foam roller so that it follows the line of your spine. Your buttocks and shoulders should both be in contact with the roller.

❷ With your legs straight and your feet pressed firmly into the floor, extend your arms over your head.

DO IT RIGHT
• Keep your legs firm throughout exercise.
• Keep your buttocks and shoulders in contact with the roller throughout exercise.

❸ Raise your head, neck, and shoulders as if to do a crunch. Leave your right leg and left arm down on the floor, using your hand for support. Raise your left leg and right arm, and reach for your ankle.

❹ Slowly roll down the roller, dropping your raised arm and leg. Repeat on the opposite leg and arm. Repeat 15 times on each side.

TARGETS
• Upper arms
• Shoulder stabilizers
• Abdominals
• Hamstrings

LEVEL
• Advanced

BENEFITS
• Improves core, pelvic, and shoulder stability

NOT ADVISABLE IF YOU HAVE . . .
• Back pain
• Neck pain

MODIFICATION

Harder: Keep one leg on the floor for support, and reach both arms toward the raised leg as you crunch up.

BEST FOR

- rectus abdominis
- transversus abdominis
- triceps brachii
- trapezius
- pectoralis major
- deltoideus anterior
- serratus anterior
- rectus femoris
- vastus intermedius
- biceps femoris
- semitendinosus
- semimembranosus

ANNOTATION KEY

Black text indicates target muscles

Gray text indicates other working muscles

* indicates deep muscles

AVOID
- Allowing your shoulders to lift toward your ears.
- Bending your knees.

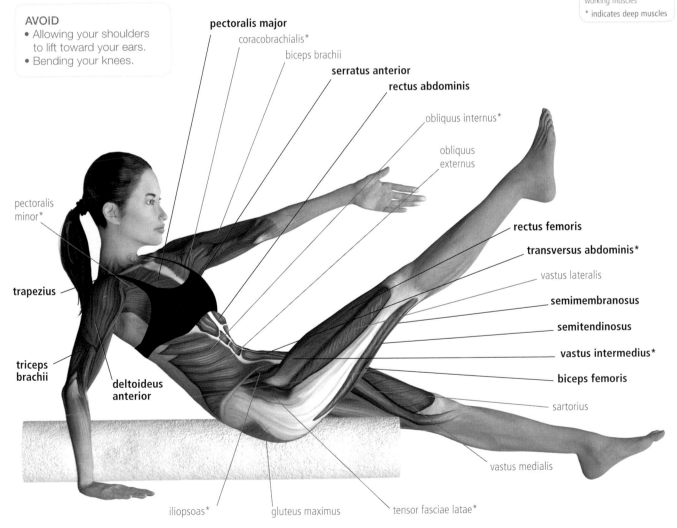

pectoralis major

coracobrachialis*

biceps brachii

serratus anterior

rectus abdominis

obliquus internus*

obliquus externus

pectoralis minor*

trapezius

triceps brachii

deltoideus anterior

rectus femoris

transversus abdominis*

vastus lateralis

semimembranosus

semitendinosus

vastus intermedius*

biceps femoris

sartorius

vastus medialis

iliopsoas*

gluteus maximus

tensor fasciae latae*

FOAM ROLLER SUPINE MARCHES

1 Lie lengthwise on a foam roller so that it follows the line of your spine. Place your arms on the floor by your sides, bending your knees so that your feet rest flat on the floor.

BEST FOR

- rectus abdominis
- transversus abdominis
- obliquus internus
- obliquus externus
- iliopsoas
- sartorius
- biceps femoris
- rectus femoris

2 Pointing your toes and keeping your hips from lifting or shifting, raise one knee toward your chest.

TARGETS
- Abdominals
- Upper arms
- Hip flexors
- Quadriceps

LEVEL
- Advanced

BENEFITS
- Improves core and pelvic stability

NOT ADVISABLE IF YOU HAVE . . .
- Lower-back pain
- Neck pain
- Shoulder pain

3 Switch legs, again being careful not to allow your hips to lift.

④ Repeat 15 times on each leg as you establish a smooth "marching" rhythm.

AVOID
- Allowing your shoulders to lift toward your ears.
- Allowing your hips and lower back to lift off the roller during the movement.

DO IT RIGHT
- Keep your legs and your toes pointed.
- Relax your neck and shoulders throughout the exercise.
- Keep your hands and forearms flat on the floor.

ANNOTATION KEY

Black text indicates target muscles

Gray text indicates other working muscles

* indicates deep muscles

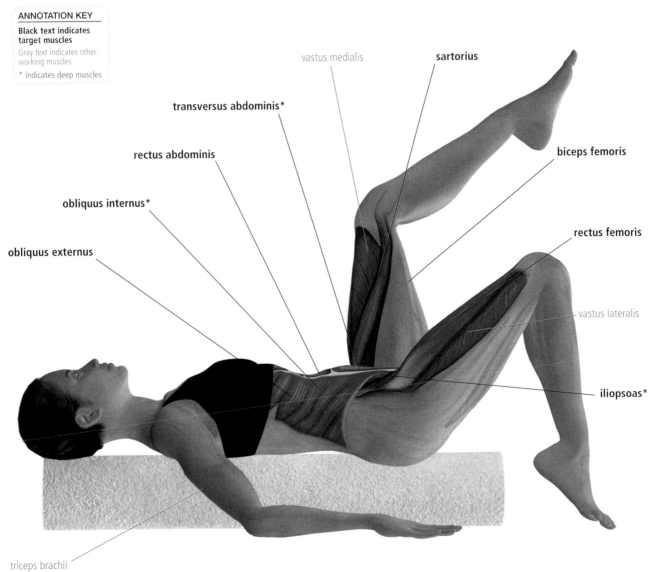

vastus medialis

sartorius

transversus abdominis*

biceps femoris

rectus abdominis

obliquus internus*

rectus femoris

obliquus externus

vastus lateralis

iliopsoas*

triceps brachii

TINY STEPS

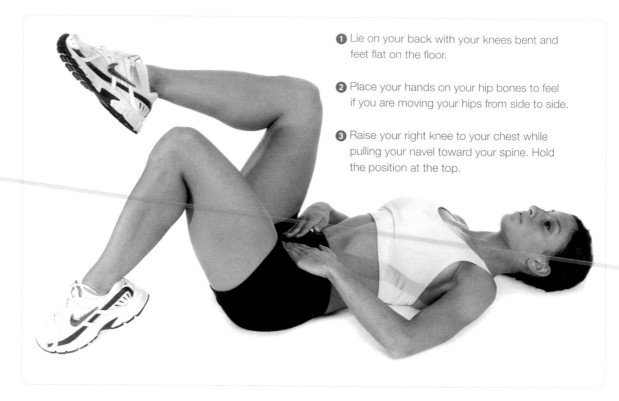

1 Lie on your back with your knees bent and feet flat on the floor.

2 Place your hands on your hip bones to feel if you are moving your hips from side to side.

3 Raise your right knee to your chest while pulling your navel toward your spine. Hold the position at the top.

TARGETS
• Lower abdominals

LEVEL
• Beginner

BENEFITS
• Develops lower-abdominal stability

NOT ADVISABLE IF YOU HAVE . . .
• Sharp lower-back pain that radiates down the legs

4 As you continue to pull your navel toward your spine, lower your right leg onto the floor while controlling any movement in your hips.

5 Alternate legs to complete the full movement. Repeat six to eight times.

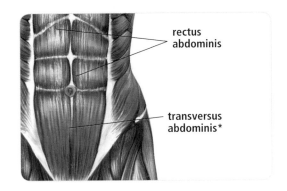

rectus abdominis

transversus abdominis*

AVOID
• Allowing your hips to move back and forth while legs are mobilized.

DO IT RIGHT
• Pull your navel in toward your spine throughout the exercise.

ANNOTATION KEY
Black text indicates target muscles
Gray text indicates other working muscles
* indicates deep muscles

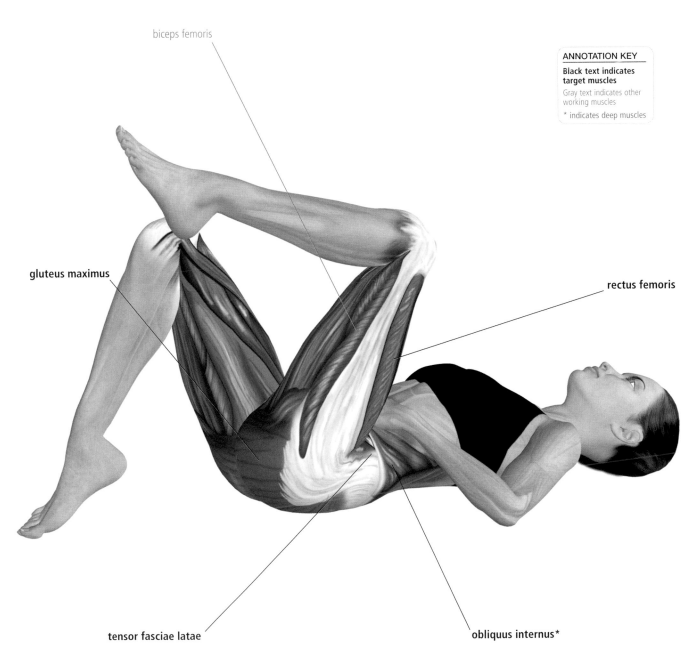

biceps femoris

gluteus maximus

rectus femoris

tensor fasciae latae

obliquus internus*

DOUBLE-LEG ABDOMINAL PRESS

1 Lie on your back with your knees and feet lifted in tabletop position, your thighs making a 90-degree angle with your upper body. Place your hands on the front of your knees, your fingers facing upward, one palm on each leg.

AVOID
• Holding your breath while performing the exercise.

TARGETS
• Total body

LEVEL
• Intermediate

BENEFITS
• Strengthens core, hip flexors, and upper arms

NOT ADVISABLE IF YOU HAVE . . .
• Back pain
• Hip pain

2 Flex your feet and, keeping your elbows bent and pulled into your sides, press your hands into your knees. Create resistance by pushing back against your hands with your knees. Hold for 1 minute, and repeat five times.

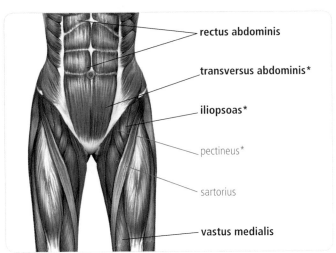

- **rectus abdominis**
- **transversus abdominis***
- **iliopsoas***
- pectineus*
- sartorius
- **vastus medialis**

BEST FOR

- rectus abdominis
- transversus abdominis
- triceps brachii
- iliopsoas
- vastus medialis
- vastus lateralis
- vastus intermedius
- rectus femoris

DO IT RIGHT

- Keep your elbows pulled in toward your sides.
- Relax your shoulders and neck.
- Flex your feet and press your knees together.
- Tuck your tailbone up toward the ceiling.

ANNOTATION KEY

Black text indicates target muscles

Gray text indicates other working muscles

* indicates deep muscles

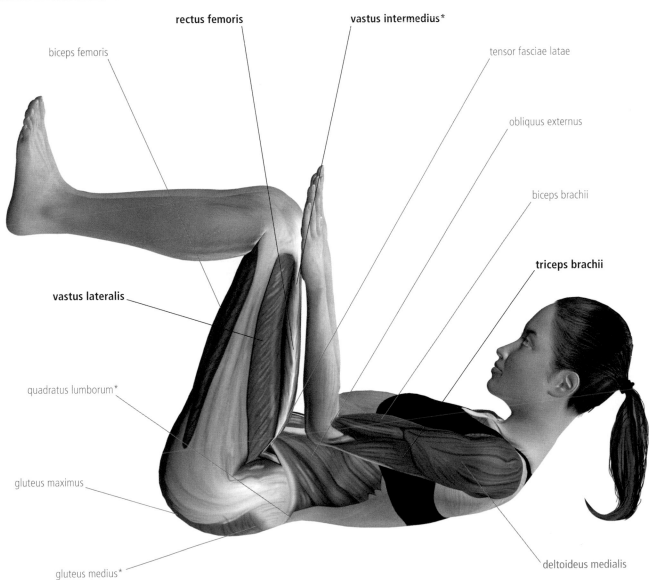

- **rectus femoris**
- **vastus intermedius***
- biceps femoris
- tensor fasciae latae
- obliquus externus
- biceps brachii
- **triceps brachii**
- **vastus lateralis**
- quadratus lumborum*
- gluteus maximus
- gluteus medius*
- deltoideus medialis

THE TWIST

1 Lie on your right side with your legs outstretched and pressed firmly together. Press your right hip into the floor, and use both hands to support your torso.

DO IT RIGHT
- Keep your limbs elongated as much as possible.
- Keep your shoulders stable.
- Lift your hips up high to reduce the weight on your upper body.

TARGETS
- Abdominals
- Shoulders

LEVEL
- Advanced

BENEFITS
- Provides a total-body workout
- Builds endurance

NOT ADVISABLE IF YOU HAVE . . .
- Shoulder issues
- Back pain
- Wrist injury

2 Position your right hand directly beneath your shoulder and press your body upward until you form a straight line from shoulder to feet.

3 Drawing your navel into your spine, extend your left arm toward the ceiling.

106

4 Bring your left arm down and across your torso, rotating the upper body to the right. Hold for a count of 10.

5 Return to the starting position, with your hip on the floor and both hands supporting your torso. Repeat sequence four to six times, and then switch sides.

BEST FOR

- latissimus dorsi
- rectus abdominis
- obliquus internus
- obliquus externus
- transversus abdominis
- adductor magnus
- adductor longus
- deltoideus medialis

AVOID

- Allowing your shoulder to sink into its socket.

ANNOTATION KEY

Black text indicates target muscles

Gray text indicates other working muscles

* indicates deep muscles

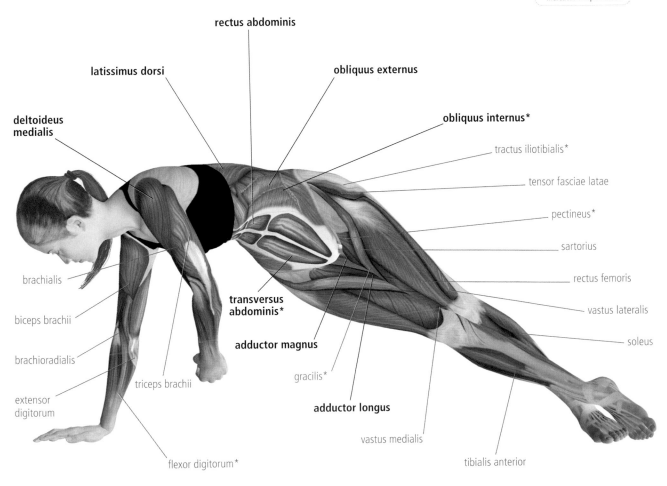

rectus abdominis

latissimus dorsi

obliquus externus

deltoideus medialis

obliquus internus*

tractus iliotibialis*

tensor fasciae latae

pectineus*

sartorius

rectus femoris

vastus lateralis

soleus

brachialis

biceps brachii

brachioradialis

extensor digitorum

triceps brachii

transversus abdominis*

adductor magnus

gracilis*

adductor longus

vastus medialis

tibialis anterior

flexor digitorum*

STANDING KNEE CRUNCH

① Stand tall with your left leg in front of the right, and extend your hands up toward the ceiling, your arms straight.

BEST FOR

- rectus abdominis
- obliquus internus
- obliquus externus
- transversus abdominis
- gluteus maximus
- gluteus medius
- tensor fasciae latae
- piriformis
- iliopsoas
- gastrocnemius
- soleus

TARGETS
- Pelvic and core stabilizers
- Abdominals
- Gluteal muscles

LEVEL
- Intermediate

BENEFITS
- Strengthens core
- Strengthens calves and gluteal muscles
- Improves balance

NOT ADVISABLE IF YOU HAVE . . .
- Knee pain

② Shift your weight onto your left foot, and raise your right knee to the height of your hips. Simultaneously go up on the toes of your left leg, while pulling your elbows down by your sides, your hands making fists. This creates the crunch.

③ Pause at the top of the movement, and then return to the starting position. Repeat the sequence with your right leg as the standing leg. Repeat 10 times on each leg.

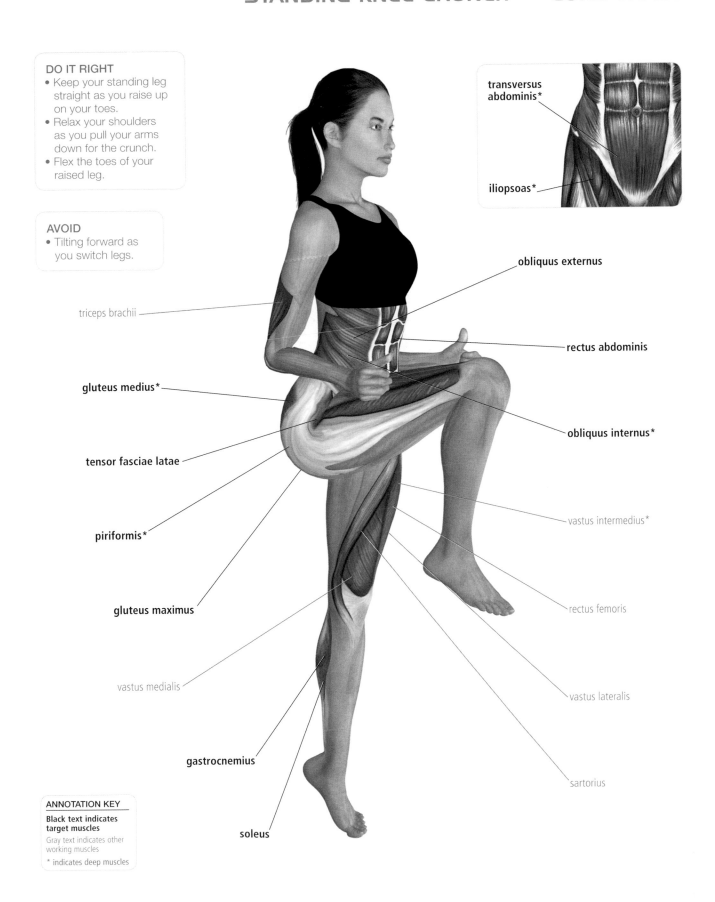

DO IT RIGHT
- Keep your standing leg straight as you raise up on your toes.
- Relax your shoulders as you pull your arms down for the crunch.
- Flex the toes of your raised leg.

AVOID
- Tilting forward as you switch legs.

transversus abdominis*

iliopsoas*

triceps brachii

obliquus externus

rectus abdominis

gluteus medius*

obliquus internus*

tensor fasciae latae

vastus intermedius*

piriformis*

gluteus maximus

rectus femoris

vastus medialis

vastus lateralis

gastrocnemius

sartorius

soleus

ANNOTATION KEY
Black text indicates target muscles
Gray text indicates other working muscles
* indicates deep muscles

POWER SQUAT

❶ Stand straight, holding a weighted medicine ball in front of your torso.

❷ Shift your weight to your left foot, and bend your right knee, lifting your right foot toward your buttocks. Bend your elbows and draw the ball toward the outside of your right ear.

AVOID
- Allowing your knee to extend beyond your toes as you bend and rotate.
- Moving your foot from its starting position.
- Flexing your spine.

❸ Maintaining a neutral spine, bend at your hips and knee. Lower your torso toward your left side, bringing the ball toward your right ankle.

❹ Press into your left leg and straighten your knee and torso, returning to the starting position. Repeat 15 times for two sets on each leg.

TARGETS
- Abdominals
- Hip flexors

LEVEL
- Advanced

BENEFITS
- Improves balance
- Stabilizes pelvis, trunk, and knees
- Promotes stronger movement patterns

NOT ADVISABLE IF YOU HAVE . . .
- Knee pain
- Lower-back pain
- Shoulder pain

DO IT RIGHT
- Move the ball in an arc through the air.
- Keep your hips and knees aligned throughout the movement.
- Relax your neck and shoulders.

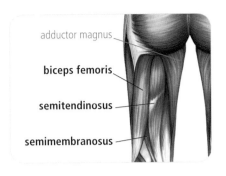

adductor magnus

biceps femoris

semitendinosus

semimembranosus

- semitendinosus
- semimembranosus
- biceps femoris
- vastus medialis
- vastus lateralis
- rectus femoris
- gluteus maximus
- gluteus medius
- piriformis
- erector spinae
- tibialis anterior
- tibialis posterior
- soleus
- gastrocnemius
- deltoideus medialis
- infraspinatus
- supraspinatus
- teres minor

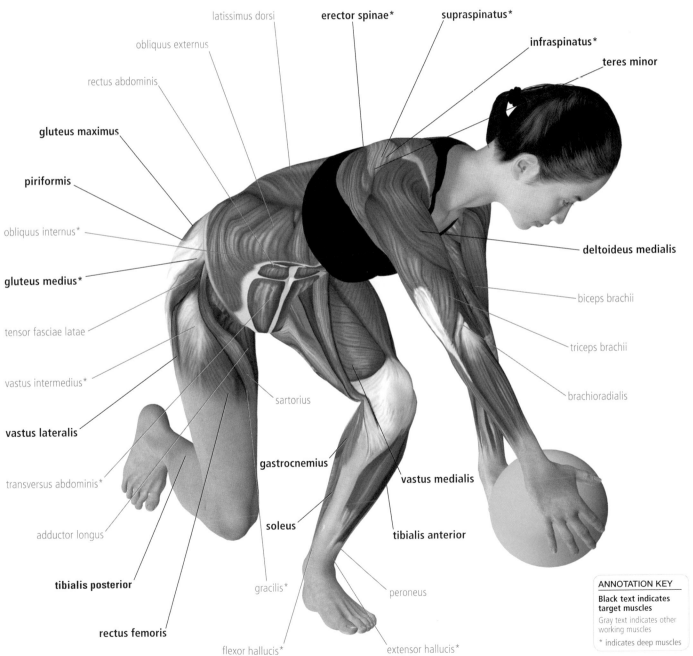

latissimus dorsi

obliquus externus

rectus abdominis

gluteus maximus

piriformis

obliquus internus*

gluteus medius*

tensor fasciae latae

vastus intermedius*

vastus lateralis

transversus abdominis*

adductor longus

tibialis posterior

gracilis*

rectus femoris

flexor hallucis*

sartorius

gastrocnemius

soleus

extensor hallucis*

peroneus

tibialis anterior

vastus medialis

erector spinae*

supraspinatus*

infraspinatus*

teres minor

deltoideus medialis

biceps brachii

triceps brachii

brachioradialis

ANNOTATION KEY

Black text indicates target muscles

Gray text indicates other working muscles

* indicates deep muscles

SWISS BALL REVERSE BRIDGE ROTATION

1 Lie with your shoulders and lower back on a Swiss ball and your feet hip-width apart. Your knees should be bent at 90 degrees.

2 Grasp a medicine ball with both hands, and position your arms straight up.

3 Rotate your upper body to the left, rolling onto your left shoulder on top of the Swiss ball.

4 Hold for 5 seconds, and then slowly roll back to the starting position with the Swiss ball in the center of your shoulders.

TARGETS
• Obliques
• Abdominals

LEVEL
• Intermediate

BENEFITS
• Stabilizes core
• Strengthens obliques and abdominals

NOT ADVISABLE IF YOU HAVE . . .
• Neck issues
• Lower-back pain

5 Repeat the exercise, rotating your torso and rolling your shoulders to the right.

AVOID
• Bending your arms.
• Continuing to rotate the Swiss ball when it is lying directly under one shoulder and both shoulders are vertical.

DO IT RIGHT

- Position the Swiss ball directly between your shoulder blades to start the exercise.
- Activate your abdominals so that you maintain neutral alignment.
- Keep your hips in line with your knees as you rotate your upper body, to work your spinal rotators.

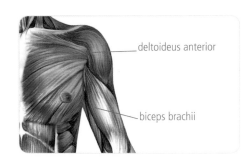

deltoideus anterior

biceps brachii

BEST FOR

- obliquus externus
- obliquus internus

ANNOTATION KEY

Black text indicates target muscles

Gray text indicates other working muscles

* indicates deep muscles

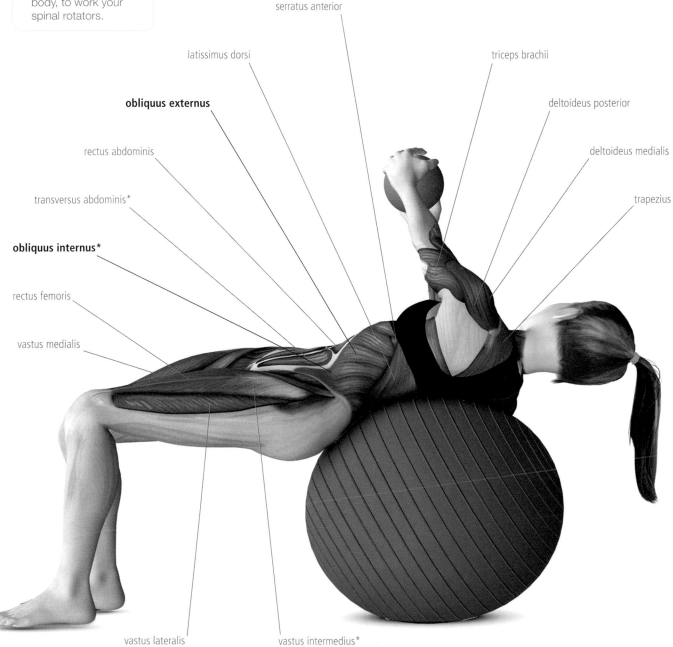

serratus anterior

latissimus dorsi

triceps brachii

obliquus externus

deltoideus posterior

rectus abdominis

deltoideus medialis

transversus abdominis*

trapezius

obliquus internus*

rectus femoris

vastus medialis

vastus lateralis

vastus intermedius*

SWISS BALL SITTING BALANCE

1 Sit on a Swiss ball with your feet together and your hands resting on the ball at your sides.

2 Lift one foot off the floor, and hold for 5 seconds.

TARGETS
- Abdominals
- Quadriceps

LEVEL
- Beginner

BENEFITS
- Stabilizes core
- Strengthens abdominals

NOT ADVISABLE IF YOU HAVE . . .
- Neck issues
- Lower-back pain

3 Put your foot down, and then lift your other foot.

4 Repeat five times on each leg.

DO IT RIGHT
- Sit up straight, and keep your abdominals activated.

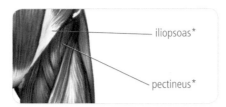

iliopsoas*

pectineus*

BEST FOR

- rectus abdominis
- transversus abdominis
- rectus femoris
- vastus lateralis
- vastus intermedius
- vastus medialis

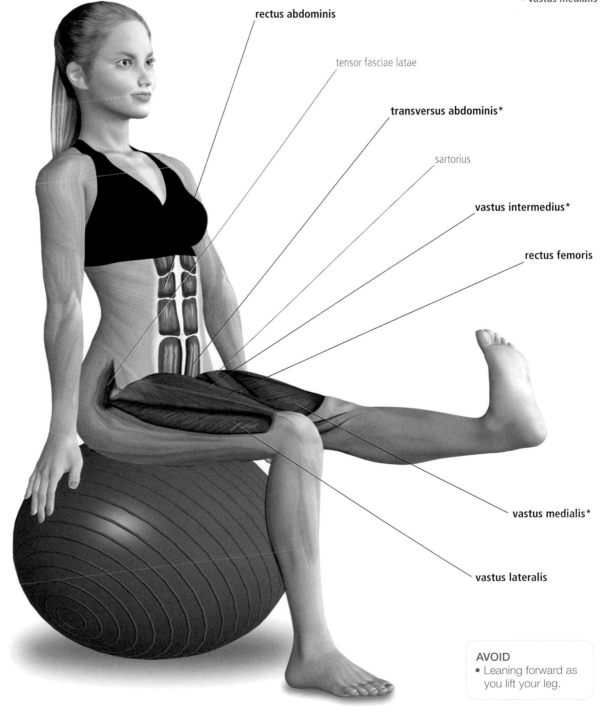

rectus abdominis

tensor fasciae latae

transversus abdominis*

sartorius

vastus intermedius*

rectus femoris

vastus medialis*

vastus lateralis

AVOID
- Leaning forward as you lift your leg.

SWISS BALL HIP CIRCLES

1 Sit on a Swiss ball with your feet together and your hands on your hips.

DO IT RIGHT
• Keep your circles small—if your feel a crunching in your neck, you are moving too widely.

AVOID
• Using your legs to initiate the movement.

2 Tighten your abdominal muscles, and use your pelvis to rotate the ball slowly to the right in small circles.

TARGETS
• Lower back
• Hips

LEVEL
• Beginner

BENEFITS
• Stabilizes core
• Stretches lower back

NOT ADVISABLE IF YOU HAVE . . .
• Lower-back pain

3 Return to the starting position, and repeat on the other side.

BEST FOR

- erector spinae
- multifidus spinae
- transversus abdominis
- obliquus externus
- quadratus lumborum
- infraspinatus
- gluteus medius
- iliopsoas

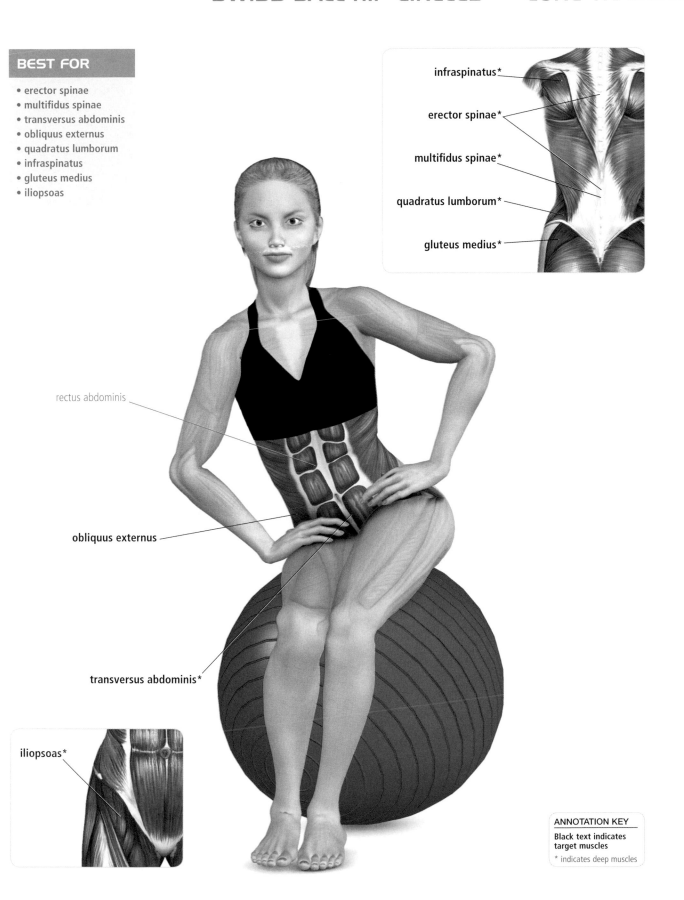

infraspinatus*

erector spinae*

multifidus spinae*

quadratus lumborum*

gluteus medius*

rectus abdominis

obliquus externus

transversus abdominis*

iliopsoas*

ANNOTATION KEY

Black text indicates target muscles

* indicates deep muscles

SWISS BALL REVERSE BRIDGE ROLL

❶ Lie with your lower back on a Swiss ball and your feet together. Your knees should be bent at 90 degrees. Position your arms out to the sides.

DO IT RIGHT
- Exhale as you roll on the ball, and inhale as you return to the starting position.
- Hold your body stable as you roll across the ball, working against the ball's natural rotation.
- Increase the space between your feet if necessary to maintain your balance.

❷ Move your upper body across the ball to the left, rolling the ball under your shoulders and toward your left shoulder.

❸ Hold for 5 seconds, and then slowly roll the ball back to the center of your shoulders.

❹ Return to the starting position, and then roll to the right. Repeat five times on each side.

TARGETS
- Obliques
- Abdominals

LEVEL
- Intermediate

BENEFITS
- Stabilizes core
- Strengthens obliques and abdominals

NOT ADVISABLE IF YOU HAVE . . .
- Neck pain
- Lower-back pain

AVOID
- Allowing your pelvis to drop out of alignment— your body should form a straight line from your shoulders to your knees.
- Continuing to rotate the ball when it is lying directly under one shoulder and both shoulders are vertical.

MODIFICATION

Easier: Rather than keeping your feet together, position them about shoulder-width apart. Then follow steps 2 through 4.

BEST FOR

- rectus abdominis
- transversus abdominis
- obliquus externus
- obliquus internus

ANNOTATION KEY
Black text indicates **target muscles**
Gray text indicates other working muscles
* indicates deep muscles

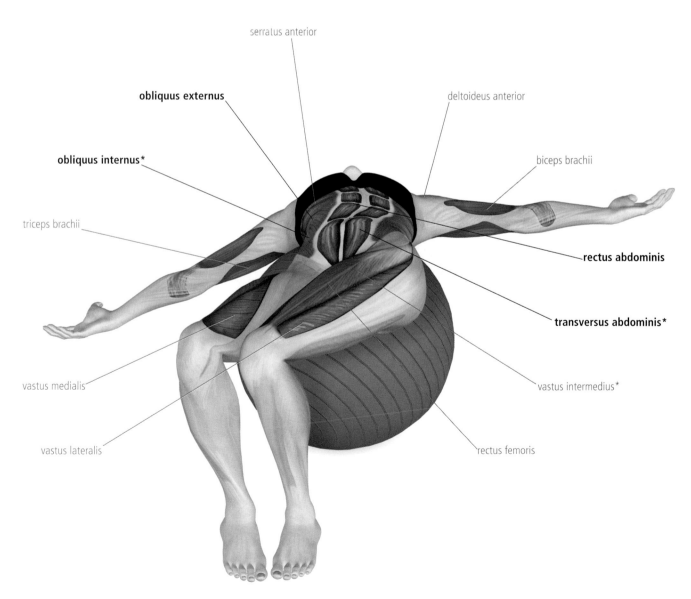

serratus anterior

obliquus externus

deltoideus anterior

obliquus internus*

biceps brachii

triceps brachii

rectus abdominis

transversus abdominis*

vastus medialis

vastus intermedius*

vastus lateralis

rectus femoris

ABDOMINAL HIP LIFT

1 Lie down with your legs in the air and crossed at the ankles, knees straight. Place your arms on the floor, straight by your sides.

DO IT RIGHT
- Keep your legs straight and firm throughout the exercise.
- Relax your neck and shoulders as you lift the hips.

AVOID
- Jerking your movements or using momentum to lift your hips.

BEST FOR
- rectus abdominis
- transversus abdominis
- vastus intermedius
- tensor fasciae latae
- gluteus maximus
- gluteus medius
- triceps brachii
- rectus femoris
- iliopsoas

TARGETS
- Abdominals
- Upper arms

LEVEL
- Intermediate

BENEFITS
- Strengthens core and pelvic stabilizers
- Firms and tones lower abdorninals

NOT ADVISABLE IF YOU HAVE . . .
- Back pain
- Neck pain
- Shoulder pain

2 Pinching your legs together and squeezing your buttocks, press into the back of your arms to lift your hips upward.

3 Slowly return your hips to the floor. Repeat 10 times, then switch with the opposite leg crossed in the front.

quadratus lumborum*

gluteus medius*

piriformis*

gluteus maximus

ANNOTATION KEY

Black text indicates target muscles

Gray text indicates other working muscles

* indicates deep muscles

MODIFICATION

Harder: Keeping your hips on the floor, raise your arms toward the ceiling. Reach toward your toes as you lift your shoulders off the floor.

rectus femoris

iliopsoas*

obliquus externus

obliquus internus*

triceps brachii

transversus abdominis*

vastus intermedius*

tensor fasciae latae

rectus abdominis

LEG RAISE

① Lie on your back with your arms along your sides. Extend your legs and lift them off the floor, angled away from your body.

AVOID
- Relying on momentum as you lift and lower your legs.
- Using your lower back to drive the movement.
- Bending your legs.

TARGETS
- Lower abdominals

LEVEL
- Intermediate

BENEFITS
- Strengthens and tones abdominals

NOT ADVISABLE IF YOU HAVE . . .
- Neck issues
- Lower-back pain

② Raise your legs until they are perpendicular to the floor.

③ Lower your legs so that your feet are just above the floor, and then raise them again, performing two sets of 20.

ANNOTATION KEY

**Black text indicates
target muscles**

Gray text indicates other
working muscles

* indicates deep muscles

DO IT RIGHT
- Keep your upper body braced.
- Use your abs to drive the movement.
- Move your legs together, as if they were a single leg.
- Keep your arms on the floor.

BEST FOR
- **rectus abdominis**
- **transversus abdominis**

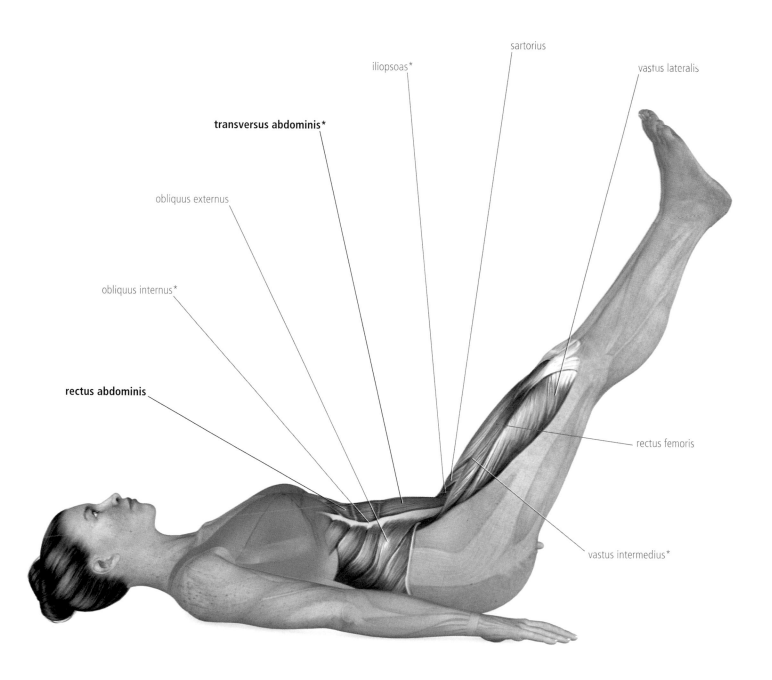

sartorius

iliopsoas*

vastus lateralis

transversus abdominis*

obliquus externus

obliquus internus*

rectus abdominis

rectus femoris

vastus intermedius*

LOWER-BODY EXERCISES

If you're like most women, a major concern is working on your lower body— trying to control those "thunder thighs" and spreading bottom. But working on your lower body doesn't just mean fitting into a pencil skirt—a conditioned lower body and strong legs can take stress off the lower back—a real bonus if you spend long days on your feet. Your gluteal muscles control your legs, hips, and pelvis, and toned glutes help protect you from lower-back and lower-limb injuries. Strong thighs (the quadriceps in the front and the hamstrings at the back) allow you to walk, run, jump, and squat. The major calf muscles are the gastrocnemius and the soleus. The gastrocnemius has two heads that when fully developed form a distinctive diamond shape (and give you a great set of gams).

LOWER BODY

1. Lie on your left side, with the foam roller placed under the middle of your thigh. Support your torso with your left forearm on the floor.

2. Bend your left leg and cross it in front of your right, so that your knee is pointed upward. Place your left foot flat on the floor.

TARGETS
- Iliotibial band
- Lateral thigh muscles
- Scapular stabilizers

LEVEL
- Intermediate

BENEFITS
- Releases the iliotibial band—this may be uncomfortable at first, but will become easier with repetition
- Strengthens the scapular stabilizers and lateral trunk muscles

NOT ADVISABLE IF YOU HAVE . . .
- Shoulder pain
- Back pain

3. Pulling with your shoulder and pushing with your supporting leg, roll back and forth along the side of your thigh. Adjust the placement of your arm as you make your motion bigger.

4. Repeat 15 times on each side.

DO IT RIGHT
- Relax your shoulders throughout the exercise.
- Press your hands and forearms firmly into the floor.

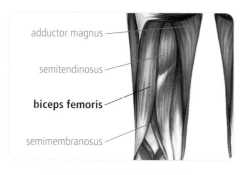

adductor magnus

semitendinosus

biceps femoris

semimembranosus

AVOID
• Allowing your shoulders to lift toward your ears.

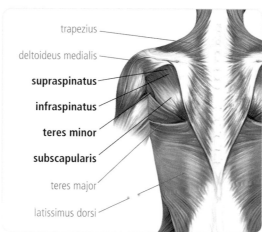

trapezius

deltoideus medialis

supraspinatus

infraspinatus

teres minor

subscapularis

teres major

latissimus dorsi

BEST FOR
• **tractus iliotibialis**
• **rectus femoris**
• **vastus medialis**
• **vastus intermedius**
• **vastus lateralis**
• **biceps femoris**
• **infraspinatus**
• **supraspinatus**
• **teres minor**
• **subscapularis**

ANNOTATION KEY
Black text indicates target muscles
Gray text indicates other working muscles
* indicates deep muscles

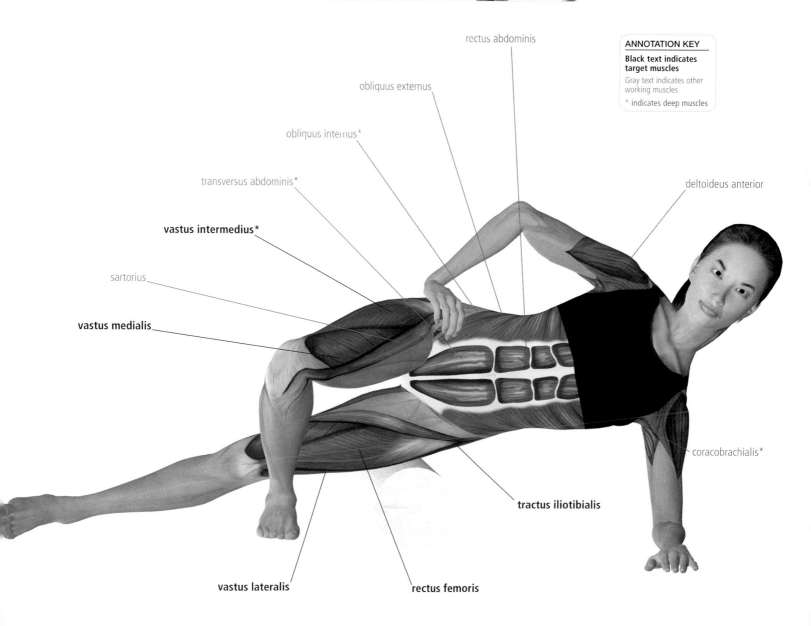

rectus abdominis

obliquus externus

obliquus internus*

transversus abdominis*

vastus intermedius*

sartorius

vastus medialis

deltoideus anterior

coracobrachialis*

tractus iliotibialis

vastus lateralis

rectus femoris

127

SWISS BALL JACKKNIFE

1 Place your hands on the floor, with your legs extended so that the tops of your feet rest on top of a Swiss ball. Keep your spine in a neutral position.

DO IT RIGHT
- Keep your chest high and retracted.
- Elongate your neck and extend your elbows throughout the movement.
- Position your hands on the floor so that they are directly below your shoulders.

2 Flex your hips, and pull your knees in toward your chest, driving your hips upward and retracting your abdomen.

TARGETS
- Abdominals
- Hip flexors

LEVEL
- Advanced

BENEFITS
- Stabilizes core
- Strengthens abdominals
- Hip flexors

NOT ADVISABLE IF YOU HAVE . . .
- Neck issues
- Lower-back pain

3 Continue to pull in until your buttocks are resting on top of your heels.

4 Hold for 5 seconds, and then extend your hips to straighten your legs and return to the starting position.

5 Repeat entire sequence three times.

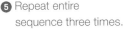

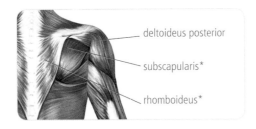

deltoideus posterior

subscapularis*

rhomboideus*

AVOID
- Bending your elbows.
- Allowing your shoulders to elevate toward your ears.

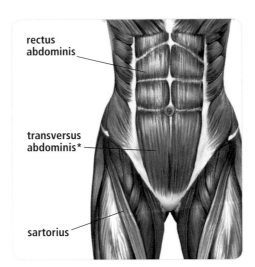

rectus abdominis

transversus abdominis*

sartorius

BEST FOR
- iliopsoas
- obliquus externus
- obliquus internus
- rectus abdominis
- sartorius
- tibialis anterior
- transversus abdominis

ANNOTATION KEY

Black text indicates target muscles

Gray text indicates other working muscles

* indicates deep muscles

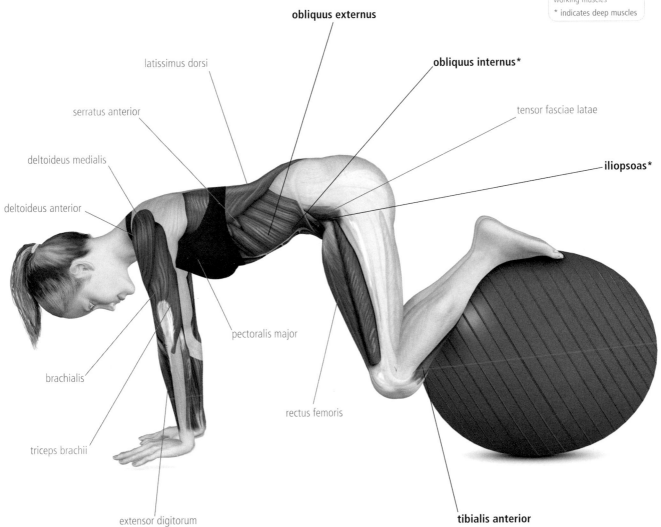

obliquus externus

latissimus dorsi

serratus anterior

deltoideus medialis

deltoideus anterior

brachialis

triceps brachii

pectoralis major

rectus femoris

extensor digitorum

obliquus internus*

tensor fasciae latae

iliopsoas*

tibialis anterior

SHOULDER BRIDGE

❶ Lie on your back with your legs bent, your feet flat on the floor, and your arms extended at your sides, angled slightly away from the body.

DO IT RIGHT
• Push through your heels, not your toes.
• Keep your knees and feet aligned.
• Keep your arms and feet on the floor.

TARGETS
• Gluteal muscles
• Hamstrings
• Quadriceps
• Lower back
• Hips

LEVEL
• Beginner

BENEFITS
• Strengthens glutes, quadriceps, and hamstrings
• Stabilizes core

NOT ADVISABLE IF YOU HAVE . . .
• Shoulder injury
• Neck injury
• Back injury

❷ Push through your heels while raising your glutes off the floor. With your feet and thighs parallel, push your arms into the floor, while extending through your fingertips.

❸ Lengthen your neck away from your shoulders as you lift your hips higher so that you form a straight line from shoulder to knee.

❹ Hold for 30 seconds to 1 minute. Exhale to release your spine slowly to the floor. Repeat three times.

MODIFICATION

Harder: Follow step 1 through 3, and then keeping your legs bent, bring your left knee toward your chest. Hold for 15 seconds, and then repeat on the other side.

MODIFICATION

Harder: Resting your feet on a foam roller, follow steps 2 and 3, and then elevate your right leg. Hold for 15 seconds, and then repeat on the other side.

AVOID
• Overextending your abdominals past your thighs in the finished position.
• Arching your back.

iliopsoas*

sartorius

vastus intermedius*

vastus medialis

BEST FOR
• erector spinae
• iliopsoas
• sartorius
• rectus femoris
• gluteus maximus
• gluteus medius
• gluteus minimus
• vastus lateralis
• vastus intermedius
• vastus medialis

ANNOTATION KEY
Black text indicates target muscles
Gray text indicates other working muscles
* indicates deep muscles

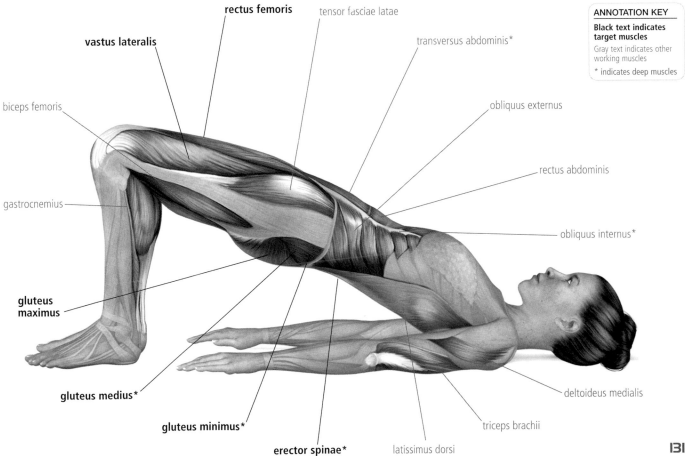

rectus femoris

tensor fasciae latae

transversus abdominis*

vastus lateralis

biceps femoris

obliquus externus

gastrocnemius

rectus abdominis

obliquus internus*

gluteus maximus

deltoideus medialis

gluteus medius*

triceps brachii

gluteus minimus*

erector spinae*

latissimus dorsi

FOAM ROLLER BICYCLE

❶ Lie on your back with a foam roller placed lengthwise under your spine, your buttocks and shoulders resting on the roller. Place your forearms on the floor on either side of the roller to balance yourself.

DO IT RIGHT
- Relax your neck throughout the exercise.
- Fully extend your leg during the downward phase of the "pedaling" movement.

❷ Draw your knees up to a tabletop position, forming a 90-degree angle between your hips, thighs, and calves.

TARGETS
- Abdominals
- Quadriceps

LEVEL
- Advanced

BENEFITS
- Improves pelvic stabilization
- Strengthens abdominals

NOT ADVISABLE IF YOU HAVE . . .
- Lower-back pain
- Neck pain

❸ Keeping your back flat, lift your head, neck, and shoulders off the roller. Straighten your right leg and pull your left knee in toward your chest, keeping your head, neck, and shoulders lifted.

AVOID
- Allowing your shoulders to lift toward your ears.
- Lifting your hips and lower back during the movement.

4 Switch legs while maintaining your balance, imitating the pedaling of a bicycle. Repeat 15 times on each leg.

BEST FOR

- rectus abdominis
- transversus abdominis
- obliquus internus
- obliquus externus
- triceps brachii
- vastus intermedius
- rectus femoris
- vastus medialis

ANNOTATION KEY

Black text indicates target muscles

Gray text indicates other working muscles

* indicates deep muscles

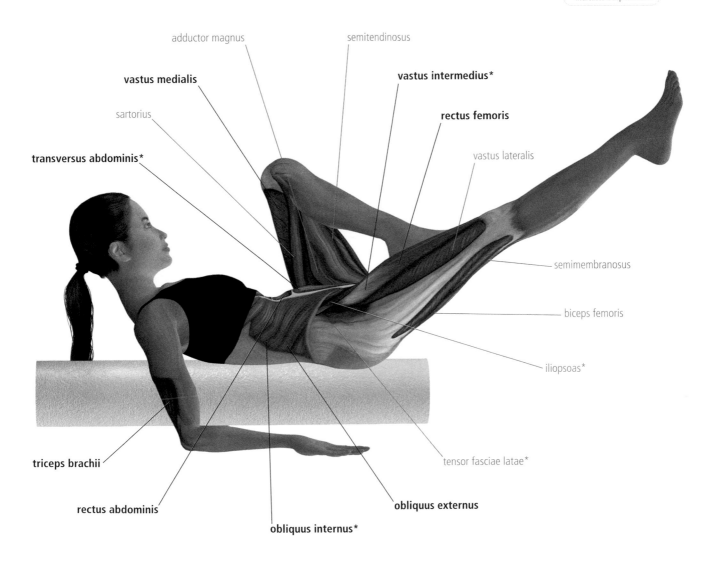

adductor magnus

semitendinosus

vastus medialis

vastus intermedius*

sartorius

rectus femoris

transversus abdominis*

vastus lateralis

semimembranosus

biceps femoris

iliopsoas*

triceps brachii

tensor fasciae latae*

rectus abdominis

obliquus externus

obliquus internus*

SINGLE-LEG CIRCLES

❶ Lie flat on the floor, with both legs and arms extended.

❷ Bend your right knee toward your chest, and then straighten your leg up in the air. Anchor the rest of your body to the floor, straightening both knees and pressing your shoulders back and down.

❸ Cross your raised leg up and over your body, aiming for your left shoulder. Continue making a circle with the raised leg, returning to the center. Add emphasis to the motion by pausing at the top between repetitions.

TARGETS
• Pelvic stability
• Abdominals

LEVEL
• Beginner

BENEFITS
• Lengthens leg muscles
• Strengthens deep abdominal muscles

NOT ADVISABLE IF YOU HAVE . . .
• Snapping hip syndrome—if this is an issue, reduce the size of the circles.

❹ Switch directions so that you aim your leg away from your body. Repeat with the other leg. Complete full movement five to eight times.

AVOID
• Making your leg circles too big to maintain stability.

BEST FOR

• rectus abdominis
• obliquus externus
• rectus femoris
• biceps femoris
• triceps brachii
• gluteus maximus
• adductor magnus
• vastus lateralis
• vastus medialis
• tensor fasciae latae

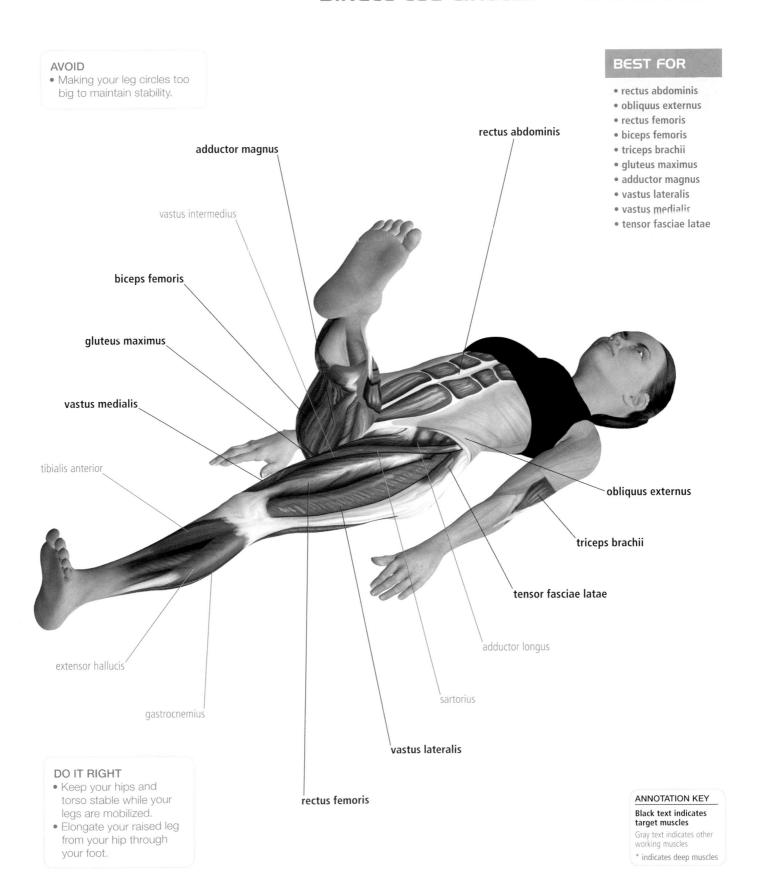

adductor magnus

rectus abdominis

vastus intermedius

biceps femoris

gluteus maximus

vastus medialis

tibialis anterior

obliquus externus

triceps brachii

tensor fasciae latae

adductor longus

extensor hallucis

sartorius

gastrocnemius

vastus lateralis

rectus femoris

DO IT RIGHT
• Keep your hips and torso stable while your legs are mobilized.
• Elongate your raised leg from your hip through your foot.

ANNOTATION KEY

Black text indicates target muscles

Gray text indicates other working muscles

* indicates deep muscles

SCISSORS

1 Lie on your back with your knees and feet lifted in tabletop position, your thighs making a 90-degree angle with your upper body, and your arms by your sides. Inhale, drawing in your abdominals.

2 Reach your legs straight up, and lift your head and shoulders off the floor. Hold the position while lengthening your legs.

DO IT RIGHT
- Keep your legs as straight as possible.
- Draw your navel into your spine.

3 Stretching your right leg away from your body, raise your left leg toward your trunk. Hold your left calf with your hands, pulsing twice while keeping your shoulders down.

TARGETS
- Quadriceps
- Abdominals

LEVEL
- Intermediate

BENEFITS
- Increases stability with unilateral movement
- Increases abdominal strength and endurance

NOT ADVISABLE IF YOU HAVE . . .
- Tight hamstrings—if this is an issue, you may bend the knee that is moving toward your chest.

AVOID
- Bending your leg.

4 Switch your legs in the air, reaching for your right leg. Stabilize your pelvis and spine. Repeat sequence six to eight times on each leg.

ANNOTATION KEY

Black text indicates target muscles
Gray text indicates other working muscles
* indicates deep muscles

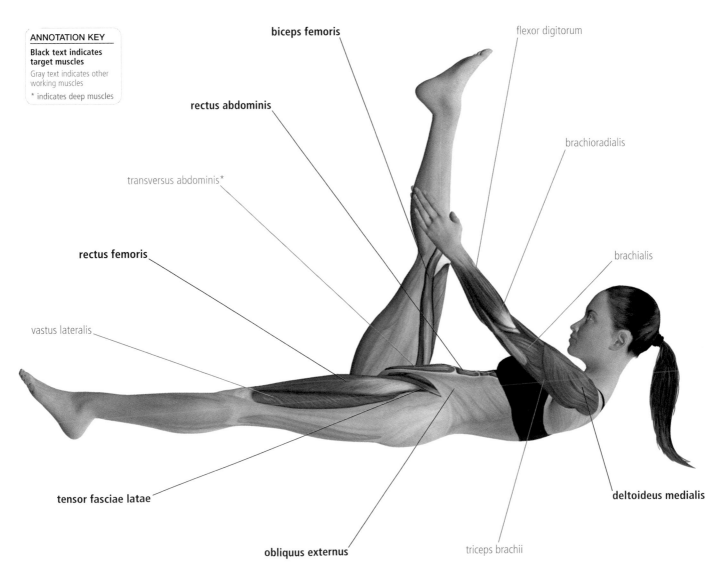

biceps femoris

rectus abdominis

transversus abdominis*

rectus femoris

vastus lateralis

tensor fasciae latae

obliquus externus

flexor digitorum

brachioradialis

brachialis

deltoideus medialis

triceps brachii

WALL SITS

DO IT RIGHT
- Keep your body firm throughout the exercise.
- Relax your shoulders and neck.
- Form a 90-degree angle with your hips and knees to receive maximum benefit from the exercise.

1 Stand with your back to a wall. Lean against the wall, and walk your feet out from under your body until your lower back rests comfortably against it.

AVOID
- Sitting below 90 degrees.
- Pushing your back into the wall to hold yourself up.
- Shifting from side to side as you begin to fatigue.

2 Glide your torso down the wall, until your hips and knees form 90-degree angles, your thighs parallel to the floor.

TARGETS
- Quadriceps
- Gluteal muscles

LEVEL
- Beginner

BENEFITS
- Strengthens quadriceps and gluteal muscles
- Trains the body to place weight evenly between the legs

3 Raise your arms straight in front of you so that they are parallel to your thighs, and relax the upper torso. Hold for 1 minute, and repeat five times.

NOT ADVISABLE IF YOU HAVE . . .
- Knee pain

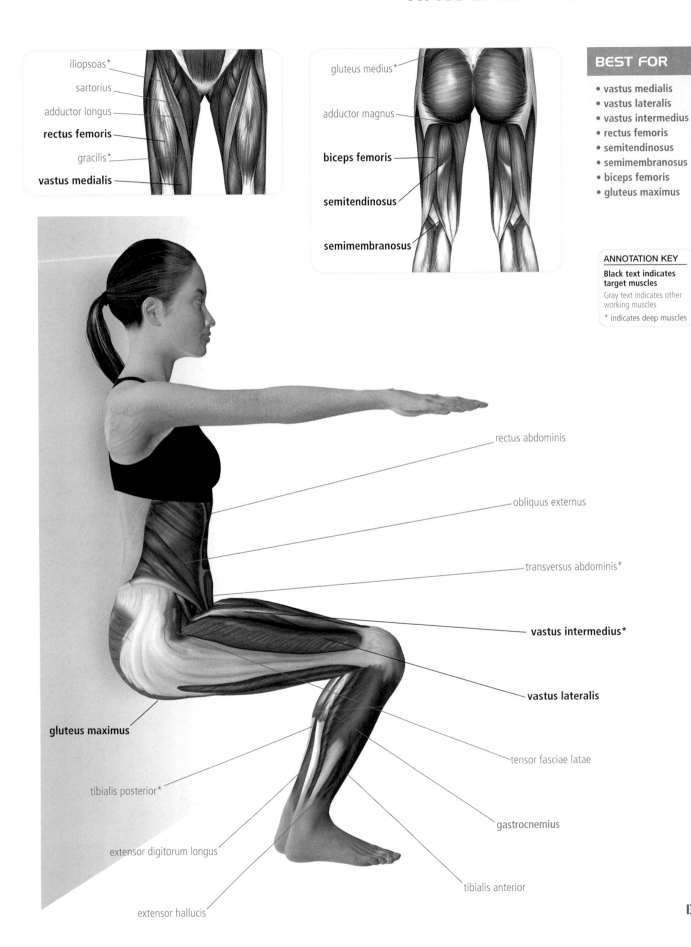

iliopsoas*
sartorius
adductor longus
rectus femoris
gracilis*
vastus medialis

gluteus medius*
adductor magnus
biceps femoris
semitendinosus
semimembranosus

BEST FOR
- vastus medialis
- vastus lateralis
- vastus intermedius
- rectus femoris
- semitendinosus
- semimembranosus
- biceps femoris
- gluteus maximus

ANNOTATION KEY
Black text indicates target muscles
Gray text indicates other working muscles
* indicates deep muscles

rectus abdominis
obliquus externus
transversus abdominis*
vastus intermedius*
vastus lateralis
tensor fasciae latae
gastrocnemius
gluteus maximus
tibialis posterior*
extensor digitorum longus
tibialis anterior
extensor hallucis

STIFF-LEGGED DEADLIFT

1 Stand upright, feet planted about shoulder-width apart, with your arms slightly in front of your thighs with a hand weight or dumbbell in each hand. Your knees should be slightly bent and your rear pushed slightly outward.

DO IT RIGHT
- Maintain the straight line of your back.
- Keep your torso stable.
- Keep your neck straight.
- Keep your arms extended.

AVOID
- Allowing your lower back to sag or arch.
- Arching your neck, straining to look forward while you are bent over.

TARGETS
- Back
- Buttocks
- Hamstrings

LEVEL
- Intermediate

BENEFITS
- Improves flexibility
- Stabilizes lower body

NOT ADVISABLE IF YOU HAVE . . .
- Lower-back pain

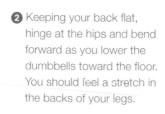

2 Keeping your back flat, hinge at the hips and bend forward as you lower the dumbbells toward the floor. You should feel a stretch in the backs of your legs.

3 With control, raise your upper body back to starting position. Repeat, completing three sets of 15.

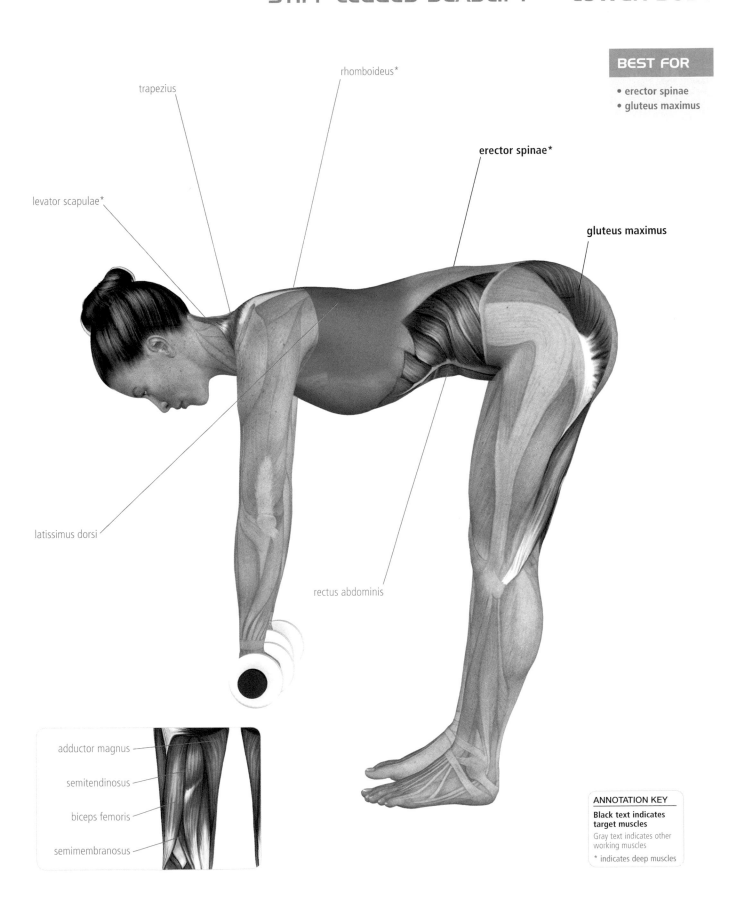

rhomboideus*

trapezius

BEST FOR

- erector spinae
- gluteus maximus

erector spinae*

levator scapulae*

gluteus maximus

latissimus dorsi

rectus abdominis

adductor magnus

semitendinosus

biceps femoris

semimembranosus

ANNOTATION KEY

Black text indicates target muscles

Gray text indicates other working muscles

* indicates deep muscles

FORWARD LUNGE

1 Stand with your feet together and your arms hanging at your sides.

2 Exhale, and carefully step back with your right leg, keeping it in line with your hips as you step back. The ball of your left foot should be in contact with the floor as you do the motion.

3 Slowly slide your right foot farther back while bending your left knee, stacking it directly above your ankle.

4 Position your palms or fingers on the floor on either side of your left leg, and slowly press your palms or fingers against the floor to enhance the placement of your upper body and your head.

5 Lift your head and gaze straight forward while leaning your upper body forward and carefully rolling your shoulders down and backward.

6 Press the ball of your right foot gradually into the floor, contract your thigh muscles, and press up to keep your left leg straight.

7 Hold for 5 seconds. Slowly return to the starting position, and then repeat on the other side.

AVOID
• Dropping your back knee to the floor.

TARGETS
• Quadriceps
• Hamstrings
• Calf muscles

LEVEL
• Beginner

BENEFITS
• Strengthens legs and arms
• Stretches groins

**NOT ADVISABLE
IF YOU HAVE . . .**
• Arm injury
• Shoulder injury
• Hip injury
• High or low blood pressure

DO IT RIGHT
• Maintain proper position of your shoulders and your whole upper body to lengthen your spine.

ANNOTATION KEY

Black text indicates target muscles

Gray text indicates other working muscles

* indicates deep muscles

BEST FOR

• biceps femoris
• adductor longus
• adductor magnus
• gastrocnemius
• tibialis posterior
• iliopsoas
• rectus femoris

gluteus medius*

pectineus*

splenius*

iliopsoas*

levator scapulae*

gluteus maximus

trapezius

tensor fasciae latae

tractus iliotibialis

vastus intermedius*

biceps femoris

gastrocnemius

vastus lateralis

soleus

plantaris

adductor magnus

tibialis posterior*

rectus femoris

flexor hallucis*

adductor longus

semitendinosus

semimembranosus

LATERAL LUNGE

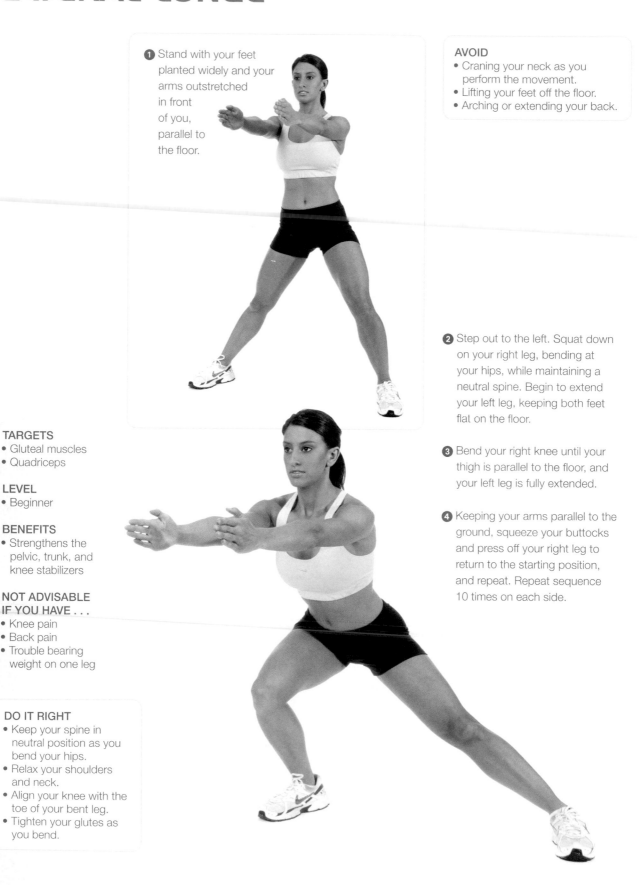

1 Stand with your feet planted widely and your arms outstretched in front of you, parallel to the floor.

AVOID
- Craning your neck as you perform the movement.
- Lifting your feet off the floor.
- Arching or extending your back.

2 Step out to the left. Squat down on your right leg, bending at your hips, while maintaining a neutral spine. Begin to extend your left leg, keeping both feet flat on the floor.

3 Bend your right knee until your thigh is parallel to the floor, and your left leg is fully extended.

4 Keeping your arms parallel to the ground, squeeze your buttocks and press off your right leg to return to the starting position, and repeat. Repeat sequence 10 times on each side.

TARGETS
- Gluteal muscles
- Quadriceps

LEVEL
- Beginner

BENEFITS
- Strengthens the pelvic, trunk, and knee stabilizers

NOT ADVISABLE IF YOU HAVE . . .
- Knee pain
- Back pain
- Trouble bearing weight on one leg

DO IT RIGHT
- Keep your spine in neutral position as you bend your hips.
- Relax your shoulders and neck.
- Align your knee with the toe of your bent leg.
- Tighten your glutes as you bend.

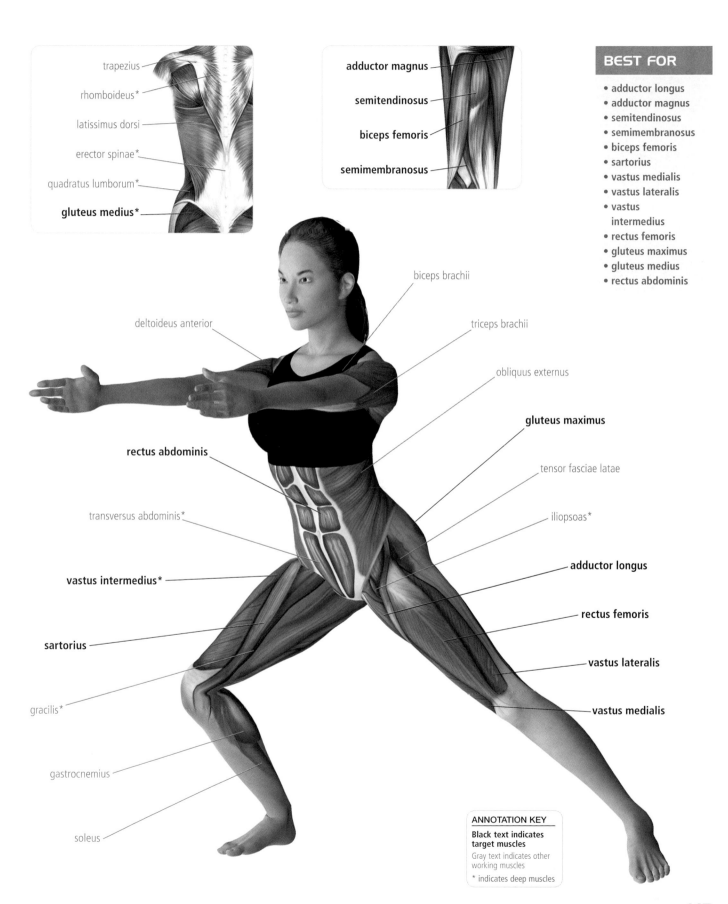

trapezius

rhomboideus*

latissimus dorsi

erector spinae*

quadratus lumborum*

gluteus medius*

adductor magnus

semitendinosus

biceps femoris

semimembranosus

biceps brachii

deltoideus anterior

triceps brachii

obliquus externus

gluteus maximus

rectus abdominis

tensor fasciae latae

transversus abdominis*

iliopsoas*

adductor longus

vastus intermedius*

rectus femoris

sartorius

vastus lateralis

gracilis*

vastus medialis

gastrocnemius

soleus

ANNOTATION KEY

Black text indicates target muscles

Gray text indicates other working muscles

* indicates deep muscles

DUMBBELL LUNGE

1 Stand with your feet planted about shoulder-width apart, with your arms at your sides and a hand weight or dumbbell in each hand.

DO IT RIGHT
- Keep your body facing forward as you step one leg in front of you.
- Stand upright.
- Gaze forward.
- Ease into the lunge.
- Make sure that your front knee is facing forward.

AVOID
- Turning your body to one side.
- Allowing your knee to extend past your foot.
- Arching your back.

TARGETS
- Gluteal muscles
- Quadriceps

LEVEL
- Intermediate

BENEFITS
- Strengthens and tones quadriceps and glutes

NOT ADVISABLE IF YOU HAVE . . .
- Knee issues

2 Keeping your head up and your spine neutral, take a big step forward.

BEST FOR

- gluteus maximus
- rectus femoris
- vastus lateralis
- vastus intermedius
- vastus medialis

3 In one movement as you step forward, bend your front knee to a 90-degree angle, and drop your front thigh until it is parallel to the floor. Your back knee will drop behind you so that you are balancing on the toe of your back foot, creating a straight line from your spine to the back of your knee.

4 Push through your front heel to stand upright, and then return to starting position. Repeat on the other leg, alternating to perform three sets of 15 lunges per leg.

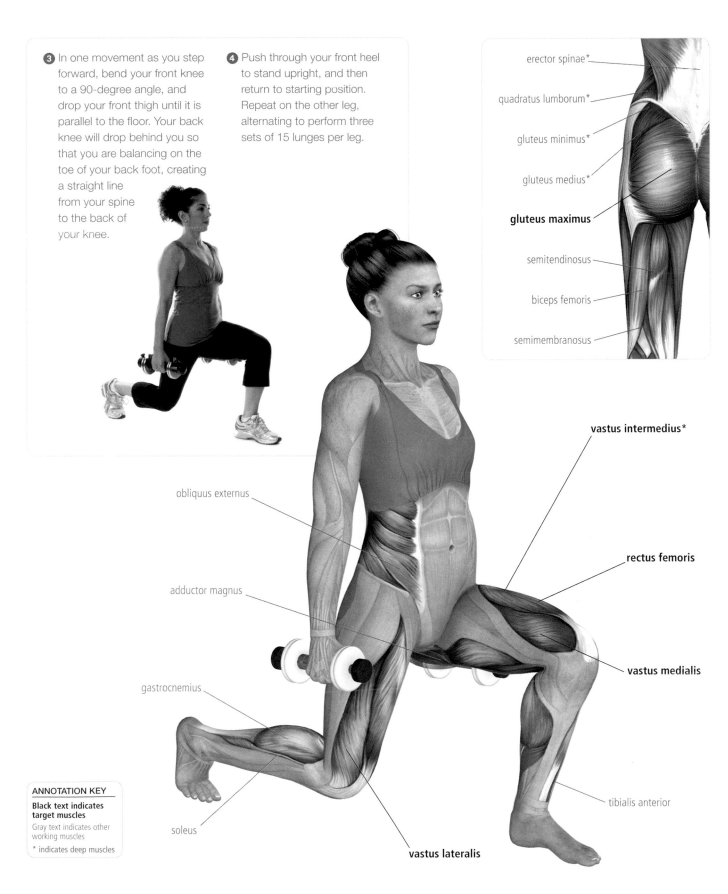

erector spinae*

quadratus lumborum*

gluteus minimus*

gluteus medius*

gluteus maximus

semitendinosus

biceps femoris

semimembranosus

obliquus externus

adductor magnus

gastrocnemius

soleus

vastus intermedius*

rectus femoris

vastus medialis

tibialis anterior

vastus lateralis

ANNOTATION KEY
Black text indicates target muscles
Gray text indicates other working muscles
* indicates deep muscles

DUMBBELL CALF RAISE

1 Stand with your arms at your sides, holding a hand weight or dumbbell in each hand with palms facing inward.

2 Keeping the rest of your body steady, slowly raise your heels off the floor to balance on the balls of your feet.

3 Hold for 10 seconds, lower, and repeat, performing three sets of 15.

TARGETS
• Calves

LEVEL
• Intermediate

BENEFITS
• Strengthens calf muscles

NOT ADVISABLE IF YOU HAVE . . .
• Ankle issues

DO IT RIGHT
• Keep your legs straight.
• Concentrate on the contraction in your calves as you balance on the balls of your feet; to feel a greater contraction, rise higher.
• Keep your core stable and your back straight.
• Gaze forward.
• Try to balance on the balls of your feet.

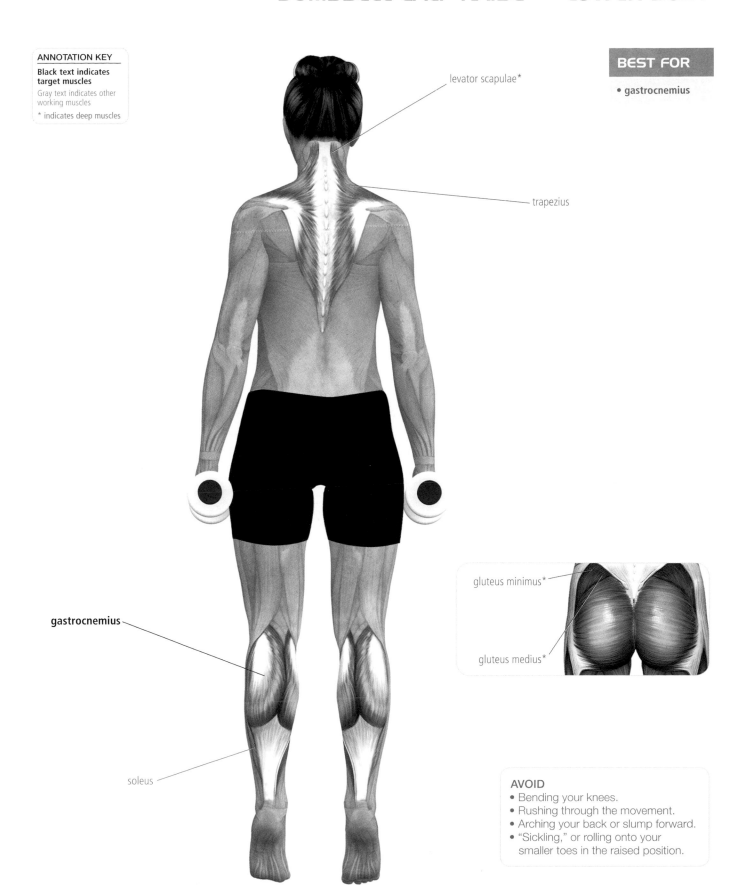

ANNOTATION KEY

Black text indicates target muscles

Gray text indicates other working muscles

* indicates deep muscles

BEST FOR

• gastrocnemius

levator scapulae*

trapezius

gluteus minimus*

gluteus medius*

gastrocnemius

soleus

AVOID
• Bending your knees.
• Rushing through the movement.
• Arching your back or slump forward.
• "Sickling," or rolling onto your smaller toes in the raised position.

KNEELING SIDE LIFT

① Kneel with your right leg outstretched to the side and your left leg lined up under your hips. Place both hands behind your head, with your elbows extended out to the sides.

AVOID
• Sinking into your neck or shoulders.

② Begin leaning your torso to the left.

TARGETS
• Abductor muscles
• Abdominals
• Gluteal muscles

LEVEL
• Advanced

BENEFITS
• Trims the waistline

NOT ADVISABLE IF YOU HAVE . . .
• Knee issues
• Back pain

③ Lift your right leg off the floor, bringing it as high as your hips. Repeat sequence five to six times. Switch sides, and repeat the sequence with your left leg.

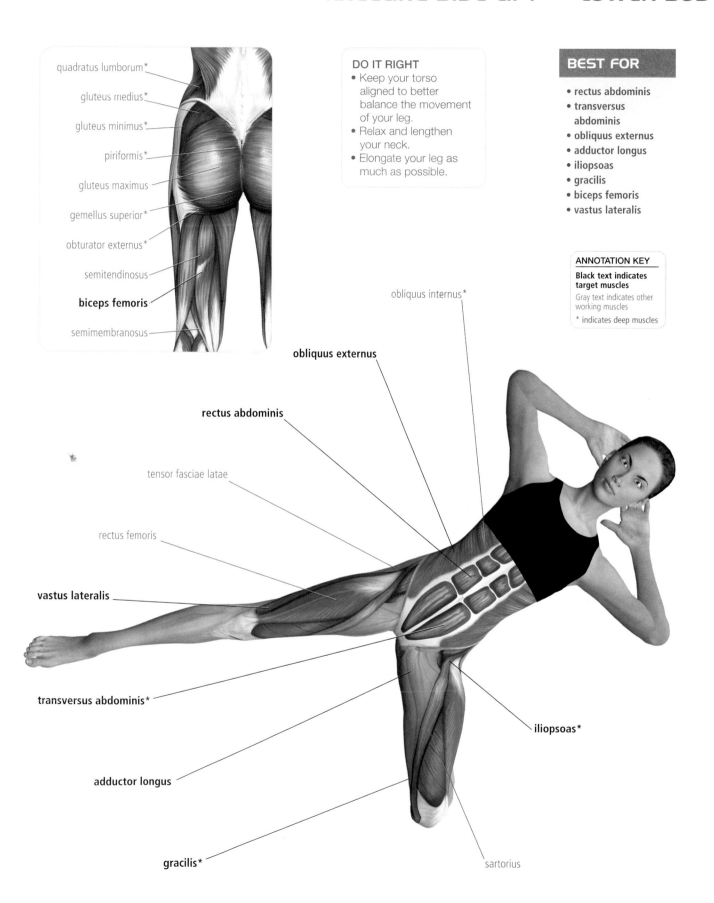

quadratus lumborum*

gluteus medius*

gluteus minimus*

piriformis*

gluteus maximus

gemellus superior*

obturator externus*

semitendinosus

biceps femoris

semimembranosus

DO IT RIGHT
- Keep your torso aligned to better balance the movement of your leg.
- Relax and lengthen your neck.
- Elongate your leg as much as possible.

BEST FOR
- rectus abdominis
- transversus abdominis
- obliquus externus
- adductor longus
- iliopsoas
- gracilis
- biceps femoris
- vastus lateralis

ANNOTATION KEY
Black text indicates target muscles
Gray text indicates other working muscles
* indicates deep muscles

obliquus internus*

obliquus externus

rectus abdominis

tensor fasciae latae

rectus femoris

vastus lateralis

transversus abdominis*

adductor longus

gracilis*

iliopsoas*

sartorius

PUT IT ALL TOGETHER:
WORKOUTS

Once you have gone through the exercises in this book and practiced executing them properly, your next step is to put these moves together into workouts. The following sequences present a sample of the many ways you can combine these exercises to target varying aims, whether you are a beginner just setting out on your first fitness regimen or a seasoned exerciser with a specific goal, such as whittling your waistline. Keep in mind that the sequence lists really just provide you with flexible frameworks—if you want to avoid a certain exercise in any one of them, simply substitute another that has a similar benefit. After trying the workouts featured here, flip through the exercises and create your own workouts to suit your individual fitness goals.

BEGINNER'S WORKOUT

A balanced introductory sequence for the exercise novice, but this workout also benefits anyone of any level.

1 Chest Stretch

page 20

2 Iliotibial Band Stretch

page 36

3 Dumbbell Upright Row

pages 66–67

4 Alternating Dumbbell Curl

pages 68–69

5 Forward Lunge

pages 142–143

6 Lateral Lunge

pages 144–145

7 Half Curl

pages 74–75

8 Swiss Ball Sitting Balance

pages 114–115

9 Swiss Ball Hip Circles

pages 116–117

10 Tiny Steps

pages 102–103

11 Shoulder Bridge

pages 130–131

12 Piriformis Stretch

page 28

BACK TO BASICS

Some things never go out of style—such as this workout featuring some tried-and-true exercise staples.

1 Triceps Stretch

page 17

2 Shoulder Bridge

pages 130–131

3 Single-Leg Circles

pages 134–135

4 Scissors

pages 136–137

5 Crunch

pages 72–73

6 Plank

pages 90–91

7 Push-Up

pages 62–63

8 Child's Pose

pages 40–41

WORKOUTS

154

LEANER LEGS, THIGHS, & GLUTES

With its focus on strengthening your abs, glutes, and legs, this workout will get you into your favorite jeans.

1 Hip-to-Thigh Stretch — page 30

2 Quadriceps Stretch — page 37

3 Standing Hamstrings Stretch — page 38

4 Standing Calf Stretch — page 39

5 Wall Sits — pages 138–139

6 Forward Lunge — pages 142–143

7 Lateral Lunge — pages 144–145

8 Dumbbell Lunge — pages 146–147

9 Dumbbell Calf Raise — pages 148–149

10 Stiff-Legged Deadlift — pages 140–141

11 Kneeling Side Lift — pages 150–151

12 Foam Roller Bicycle — pages 132–133

ARM TONER

Practice this workout regularly, and you'll soon be showing off toned arms in sleeveless tops and dresses.

1 Chair Dip — pages 44–45

2 Overhead Press — pages 48–49

3 Alternating Chest Press — pages 50–51

4 Standing Fly — pages 52–53

5 Swiss Ball Pullover — pages 56–57

6 Swiss Ball Triceps Extension — pages 58–59

7 Swiss Ball Fly — pages 60–61

8 Alternating Dumbbell Curl — pages 68–69

CORE STRENGTH & STABILITY

Working the core is one of the surest ways to get fit and strong, so you'll look and feel your best.

1 Half Curl
pages 74–75

2 The Boat
pages 84–85

3 V-Up
pages 86–87

4 Backward Ball Stretch
pages 88–89

5 Swiss Ball Transverse Abs
pages 92–93

6 Swiss Ball Rollout
pages 94–95

7 Foam Roller Calf Press
pages 96–97

8 Foam Roller Diagonal Crunch
pages 98–99

9 Foam Roller Supine Marches
pages 100–101

10 Tiny Steps
pages 102–103

11 Abdominal Hip Lift
pages 120–121

12 Leg Raise
pages 122–123

WORKING THE WAISTLINE

A focus on the obliques will help you trim and tone your midsection and take inches off your waist.

1 Seated Russian Twist
pages 76–77

2 Spine Twist
pages 78–79

3 Oblique Roll-Down
pages 80–81

4 Bicycle Crunch
pages 82–83

5 Swiss Ball Hip Crossover
pages 32–33

6 Swiss Ball Reverse Bridge Rotation
pages 112–113

7 Swiss Ball Hip Circles
pages 116–117

8 Swiss Ball Reverse Bridge Roll
pages 118–119

ALL-OVER TONING

This is a great plan for getting a full-body workout that helps you achieve maximum performance levels.

1 Posterior Hand Clasp
pages 18–19

2 Swiss Ball Kneeling Lat Stretch
page 21

3 Prone Trunk Raise
pages 64–65

4 Chair Dip
pages 44–45

5 Chair Crunch
pages 46–47

6 Double-Leg Abdominal Press
page 104–105

7 The Twist
pages 106–107

8 Standing Knee Crunch
pages 108–109

9 Power Squat
110–111

10 Foam Roller Iliotibial Band Release
pages 126–127

11 Swiss Ball Jackknife
pages 128–129

12 Upward Plank
pages 54–55

STRETCH IT OUT

Give your back a soothing stretch with this quick flexibility workout that takes less than 10 minutes.

1 Neck Side Bend
page 16

2 Latissimus Dorsi Stretch
pages 22–23

3 Toe Touch
pages 24–25

4 Cat and Dog Stretch
pages 26–27

5 Hip Stretch
page 29

6 Spine Stretch
page 31

7 Knee-to-Chest Hug
pages 34–35

8 Child's Pose
pages 40–41

GLOSSARY

GENERAL TERMS

abduction: Movement away from the body.

adduction: Movement toward the body.

anterior: Located in the front.

cardiovascular exercise: Any exercise that increases the heart rate, making oxygen and nutrient-rich blood available to working muscles.

core: Refers to the deep muscle layers that lie close to the spine and provide structural support for the entire body. The core is divisible into two groups: major core and minor core muscles. The major muscles reside on the trunk and include the belly area and the mid and lower back. This area encompasses the pelvic floor muscles (levator ani, pubococcygeus, iliococcygeus, puborectalis, and coccygeus), the abdominals (rectus abdominis, transversus abdominis, obliquus externus, and obliquus internus), the spinal extensors (multifidus spinae, erector spinae, splenius, longissimus thoracis, and semispinalis), and the diaphragm. The minor core muscles include the latissimus dorsi, gluteus maximus, and trapezius (upper, middle, and lower). Minor core muscles assist the major muscles when the body engages in activities or movements that require added stability.

crunch: A common abdominal exercise that calls for curling the shoulders toward the pelvis while lying supine with hands behind the head and knees bent.

curl: An exercise movement, usually targeting the biceps brachii, that calls for a weight to be moved through an arc, in a "curling" motion.

deadlift: An exercise movement that calls for lifting a weight, such as a dumbbell, off the floor from a stabilized bent-over position.

dumbbell: A basic piece of equipment that consists of a short bar on which plates are secured. A person can use a dumbbell in one or both hands during an exercise. Most gyms offer dumbbells with the weight plates welded on and poundage indicated on the plates, but many dumbbells intended for home use come with removable plates that allow you to adjust the weight.

extension: The act of straightening.

extensor muscle: A muscle serving to extend a body part away from the body.

flexion: The bending of a joint.

flexor muscle: A muscle that decreases the angle between two bones, as when bending the arm at the elbow or raising the thigh toward the stomach.

fly: An exercise movement in which the hand and arm move through an arc while the elbow is kept at a constant angle. Flyes work the muscles of the upper body.

hand weight: Any of a range of free weights that are often used in weight training and toning. Small hand weights are usually cast iron formed in the shape of a dumbbell, sometimes coated with rubber or neoprene for comfort.

iliotibial band (ITB): A thick band of fibrous tissue that runs down the outside of the leg, beginning at the hip and extending to the outer side of the tibia just below the knee joint. The band functions in concert with several of the thigh muscles to provide stability to the outside of the knee joint.

lateral: Located on, or extending toward, the outside.

medial: Located on, or extending toward, the middle.

medicine ball: A small weighted ball used in weight training and toning.

neutral position (spine): A spinal position resembling an S shape, consisting of a inward curve in the lower back, when viewed in profile.

posterior: Located behind.

press: An exercise movement that calls for moving a weight or other resistance away from the body.

range of motion: The distance and direction a joint can move between the flexed position and the extended position.

resistance band: Any rubber tubing or flat band device that provides a resistive force used for strength training. Also called a "fitness band," "Thera-Band," "Dyna-Band," "stretching band," and "exercise band."

rotator muscle: One of a group of muscles that assist the rotation of a joint, such as the hip or the shoulder.

scapula: The protrusion of bone on the mid to upper back, also known as the "shoulder blade."

squat: An exercise movement that calls for moving the hips back and bending the knees and hips to lower the torso and an accompanying weight, and then returning to the upright position. A squat primarily targets the muscles of the thighs, hips and buttocks, and hamstrings.

Swiss ball: A flexile, inflatable PVC ball measuring approximately 14 to 34 inches in circumference that is used for weight training, physical therapy, balance training, and many other exercise regimens. It is also called a "balance ball," "fitness ball," "stability ball," "exercise ball," "gym ball," "physioball," "body ball," and many other names.

GLOSSARY

warm-up: Any form of light exercise of short duration that prepares the body for more intense exercises.

weight: Refers to the plates or weight stacks, or the actual poundage listed on the bar or dumbbell.

LATIN TERMS

The following glossary explains the Latin terminology used to describe the muscles of the human body. Certain words are derived from Greek, which is indicated in each instance.

CHEST

coracobrachialis: Greek *korakoeidés*, "ravenlike," and *brachium*, "arm"

pectoralis (major and minor): *pectus*, "breast"

ABDOMEN

obliquus externus: *obliquus*, "slanting," and *externus*, "outward"

obliquus internus: *obliquus*, "slanting," and *internus*, "within"

rectus abdominis: *rego*, "straight, upright," and *abdomen*, "belly"

serratus anterior: *serra*, "saw," and *ante*, "before"

transversus abdominis: *transversus*, "athwart," and *abdomen*, "belly"

NECK

scalenus: Greek *skalénós*, "unequal"

semispinalis: *semi*, "half," and *spinae*, "spine"

splenius: Greek *spléníon*, "plaster, patch"

sternocleidomastoideus: Greek *stérnon*, "chest," Greek *kleís*, "key," and Greek *mastoeidés*, "breastlike"

BACK

erector spinae: *erectus*, "straight," and *spina*, "thorn"

latissimus dorsi: *latus*, "wide," and *dorsum*, "back"

multifidus spinae: *multifid*, "to cut into divisions," and *spinae*, "spine"

quadratus lumborum: *quadratus*, "square, rectangular," and *lumbus*, "loin"

rhomboideus: Greek *rhembesthai*, "to spin"

trapezius: Greek *trapezion*, "small table"

SHOULDERS

deltoideus (anterior, medial, and posterior): Greek *deltoeidés*, "delta-shaped"

infraspinatus: *infra*, "under," and *spina*, "thorn"

levator scapulae: *levare*, "to raise," and *scapulae*, "shoulder [blades]"

subscapularis: *sub*, "below," and *scapulae*, "shoulder [blades]"

supraspinatus: *supra*, "above," and *spina*, "thorn"

teres (major and minor): *teres*, "rounded"

UPPER ARM

biceps brachii: *biceps*, "two-headed," and *brachium*, "arm"

brachialis: *brachium*, "arm"

triceps brachii: *triceps*, "three-headed," and *brachium*, "arm"

LOWER ARM

anconeus: Greek *anconad*, "elbow"

brachioradialis: *brachium*, "arm," and *radius*, "spoke"

extensor carpi radialis: *extendere*, "to extend," Greek *karpós*, "wrist," and *radius*, "spoke"

extensor digitorum: *extendere*, "to extend," and *digitus*, "finger, toe"

flexor carpi pollicis longus: *flectere*, "to bend," Greek *karpós*, "wrist," *pollicis*, "thumb," and *longus*, "long"

flexor carpi radialis: *flectere*, "to bend," Greek *karpós*, "wrist," and *radius*, "spoke"

flexor carpi ulnaris: *flectere*, "to bend," Greek *karpós*, "wrist," and *ulnaris*, "forearm"

flexor digitorum: *flectere*, "to bend," and *digitus*, "finger, toe"

palmaris longus: *palmaris*, "palm," and *longus*, "long"

pronator teres: *pronate*, "to rotate," and *teres*, "rounded"

HIPS

gemellus (inferior and superior): *geminus*, "twin"

gluteus maximus: Greek *gloutós*, "rump," and *maximus*, "largest"

gluteus medius: Greek *gloutós*, "rump," and *medialis*, "middle"

gluteus minimus: Greek *gloutós*, "rump," and *minimus*, "smallest"

iliopsoas: *ilium*, "groin," and Greek *psoa*, "groin muscle"

iliacus: *ilium*, "groin"

obturator externus: *obturare*, "to block," and *externus*, "outward"

obturator internus: *obturare*, "to block," and *internus*, "within"

pectineus: *pectin*, "comb"

piriformis: *pirum*, "pear," and *forma*, "shape"

quadratus femoris: *quadratus*, "square, rectangular," and *femur*, "thigh"

UPPER LEG

adductor longus: *adducere*, "to contract," and *longus*, "long"

adductor magnus: *adducere*, "to contract," and *magnus*, "major"

biceps femoris: *biceps*, "two-headed," and *femur*, "thigh"

gracilis: *gracilis*, "slim, slender"

rectus femoris: *rego*, "straight, upright," and *femur*, "thigh"

sartorius: *sarcio*, "to patch" or "to repair"

semimembranosus: *semi*, "half," and *membrum*, "limb"

semitendinosus: *semi*, "half," and *tendo*, "tendon"

tensor fasciae latae: *tenere*, "to stretch," *fasciae*, "band," and *latae*, "laid down"

vastus intermedius: *vastus*, "immense, huge," and *intermedius*, "between"

vastus lateralis: *vastus*, "immense, huge," and lateralis, "side"

vastus medialis: *vastus*, "immense, huge," and *medialis*, "middle"

LOWER LEG

adductor digiti minimi: *adducere*, "to contract," *digitus*, "finger, toe," and *minimum* "smallest"

adductor hallucis: *adducere*, "to contract," and *hallex*, "big toe"

extensor digitorum: *extendere*, "to extend," and *digitus*, "finger, toe"

extensor hallucis: *extendere*, "to extend," and *hallex*, "big toe"

flexor digitorum: *flectere*, "to bend," and *digitus*, "finger, toe"

flexor hallucis: *flectere*, "to bend," and *hallex*, "big toe"

gastrocnemius: Greek *gastroknémía*, "calf [of the leg]"

peroneus: *peronei*, "of the fibula"

plantaris: *planta*, "the sole"

soleus: *solea*, "sandal"

tibialis anterior: *tibia*, "reed pipe," and *ante*, "before"

tibialis posterior: *tibia*, "reed pipe," and *posterus*, "coming after"

CREDITS & ACKNOWLEDGMENTS

All photographs by Jonathan Conklin/Jonathan Conklin Photography, Inc., except the following: page 11 top left Serg64/Shutterstock.com, and top middle picamaniac/Shutterstock.com.

Models: Elaine Altholz, Goldie Oren, and Melissa Grant

All large anatomical illustrations by Hector Aiza/3D Labz Animation India (www.3dlabz.com), with small insets by Linda Bucklin/Shutterstock.com

Acknowledgments

The author and publisher also offer thanks to those closely involved in the creation of this book: Moseley Road president Sean Moore; production designer Adam Moore; designers Danielle Scaramuzzo and Terasa Bernard; and photographer Jonathan Conklin.